Table of Contents

Glossary 6
Square Wave Gallery 9
(1977) Atari TIA (Television Interface Adaptor) 15
TIA Distortion Gallery 24
(1978) Intel 8244 (NTSC)/8245 (PAL) 34
(1978) Texas Instruments SN76477 37
(1978) Bally-Midway 0066-117XX 41
(1978) Signetics 2636 PVI 44
(1978) General Instrument AY-3-891x 47
Speakers, beepers, and squeakers 55
How to play back samples on something that can't play samples 58
(1979) RCA CDP1863/4 64
(1979) Williams HC-55516 67
(1979) Atari C012294 (POKEY) 69
(1979) Texas Instruments SN76489 76
(1980) Texas Instruments TMS3615N/TMS3617NS 83
(1980) National Semiconductor MM5837 85
Arcade Speech Synthesis - when it was interesting! 87
(1980) Commodore MOS 6560/6561 (VIC) 95
(1980) Namco WSG/52xx/54xx/15xx 99
WSG Family Waveforms 104
(1980) RCA CDP1869 110
Speech Synthesis - Home systems 113
(1981) OKI MSM5205 ADPCM 122
(1981) NEC µPD1771C 124
Commodore 64 Tape Loaders 128
(1982) Commodore MOS6581/6582/8580 SID 135
(1982) NEC µPD7751 150
(1983) OKI MSM5232RS 152
(1983) SNK Wave 156
(1983) NES APU – Ricoh 2A03 (NTSC)/2A07 (PAL) 158
NES/Famicom Expansions 163
Japanese vs European Chipmusic 169
(1983) CEM3394 Synthesizer Voice 175

(1984) Atari AMY 181
(1984) Yamaha YM2149F 184
(1984) Yamaha YM2151 (OPM) 190
(1984) Yamaha YM2203 (OPN) 194
(1984) Ensoniq ES5503 198
(1984) Commodore MOS7360 (TED) 202
(1984) Namco CUS30 206
(1984) Yamaha YM3526 (OPL) 209
(1985) Sega 315-5218 (SegaPCM) 214
(1985) Commodore MOS8364 (Paula) 217
(1985) DPC/Dave (Enterprise 64/128) 221
(1986) Konami K051649/K005289 SCC1 224
Super Tech Talk: Noise Generation in 8-bit Sound Chips 226
Super Tech Talk: Assigning pitches to sound chips 240

An Anthology of Sound Chips Vol. 1

© 2021 High Technology Publishing Ltd. All Rights Reserved.

Chris Abbott and Andrew Laggan have asserted the right to be identified as the authors of this work in accordance with the Copyright, Designs and Patents Act 1988.

Credits

Written and edited by Chris Abbott and Andrew Laggan.

Front cover by Toni Gálvez.

Layout and design by Ian Flory.

Thanks to Mike Clarke, David Knapp and David Youd for their technical sections and additional technical advice.

Proof-reading by Nate Lawson, Andrew Fisher, David Youd, Mike Clarke, Damian Manning, and Thomas Finnerup.

Additional material from Allister Brimble, Barry Leitch, Martin Galway, Toni Gálvez, Andrew Fisher, and Gari Biasillo.

Additional research by Russell F. Howard and Richard Karsmakers.

Executive Production by Damian Manning and Anna Black.

Dedicated to the memory of Ben Daglish, Richard Joseph, Anthony Lees, Paul Hadrill, Jason "TMR" Kelk and Paul van der Valk.

Thanks to emulator teams everywhere, especially the MAME team.

MAME was invaluable for this project. History would be lost if it wasn't for your efforts!

About this book

Content for this book was adapted for paperback from Volumes 1 and 2 of "The Little Book of Sound Chips".

The "Little Books" contain not only the chip write-ups in full colour, but beautiful QR-code/hyperlink-filled galleries of games using the chip so you can look and listen.

Find the colourful digital and hardback books here (Volume 1 is over 300 pages and volume 2 is over 400 pages).

http://c64audio.com/lbsc

Use coupon code MSM5205 for 10% discount.

Also works on other C64Audio.com holy-grail projects like "SID done by an 80-piece orchestra" (8-Bit Symphony Pro) and "Rob Hubbard does new SIDs" (Project Hubbard: Rob Returns) as well as the forthcoming Rob Hubbard biography/reference guide.

About the Authors

Chris Abbott has been fascinated by computer game music even before computer games, being entranced by Moog versions of classical pieces in his formative years.

A very brief stint as a game composer for Superior Software was followed by a period of real work until he bought an AWE32 sound card in 1994 and started sequencing Commodore 64 tunes. That grew into a 25-year journey in computer game music remixing, covering everything from studio CDs to DJ events to classical concerts, in projects such as "Back in Time", "Back in Time Live", and "8-Bit Symphony".

Chris has written articles and been interviewed for numerous magazines and websites such as Edge, Retro Gamer, Super Play and more. He has also been interviewed on BBC 1, Radio 4, and podcasts galore and is Rob Hubbard's authorised biographer.

Andrew Laggan is a graphic artist by trade, and a keen video game historian and retro enthusiast. He is also a hobbyist musician.

His love for game music stems back to the days of the Sinclair Spectrum with favourite sound chips being the MOS6581 SID and the SNES SPC700.

Glossary

There are words and phrases referred to in this book that it would be handy to quickly flesh out here!

SSG

Software-controlled Sound Generator.

This is used throughout the book to refer to chips that generate the waveforms they play from scratch, rather than using samples. Tone generators and noise generators fall into this category. The term PSG ("Programmable Sound Generator") could also apply.

Amplitude

How loud something is.

Modulation

When something changes, or is changed, over time.

Waveform

A change in amplitude over time.

FM

FM (Frequency Modulation) synthesis modifies the frequency of one waveform (known as an 'operator') very quickly using another waveform. You can then take that waveform and modify its frequency using another, and so on. The more operators you have, the more complex tones you can produce.

Wavetable

A wavetable has always been seen as a very small sample, for instance, a sound effect or looped waveform approximation.

Wavetables are commonly used as the basis for other technologies (for example, FM Synthesis), or to improve them. Pretty much every digital synthesizer uses a set of wavetables to clean up the sound of digital artefacts such as aliasing.

Wavetable Synthesis allows cross-fading between multiple wavetables over time.

PCM (Pulse Code Modulation)

The conventional method for converting analog audio into digital audio. 8-bit PCM is encoded using values between 0-255, 16-bit using values between 0-65535. Other encoding methods exist.

ADPCM

A variation of PCM encoding that compresses a sound to 4-bits (from 8-bit or 16-bit) while adding small amounts of extra filter information to help restore some of the quality lost through compression.

This was the main standard for video game sampled audio for decades.

Trackers/MODs

Essentially "composing by spreadsheet", music is composed by creating short sequences of notes and commands known as "patterns".

These patterns are used to construct "orderlists", which show the order in which to play the patterns, and how to play them (tempo changes, transpositions, sample assignments, etc.).

Trackers produce various self-contained playable files, the most popular of which is the MOD(ule) format, and later, the XM (eXtended Module) format.

DAC

Digital-to-analog converter. For audio, we need DACs to turn digital representation of sound into a wave that your ears can pick up. DACs can be of varying quality, their specifications having a huge impact on overall sound quality.

About frequencies and tuning

You might have wondered why some chips occasionally sound out-of-tune. Well, as a matter of fact, all sound chips are out of tune. The key question is, by how much?

To accurately represent the frequencies of any standard musical scale you need very accurate numbers (for example, in the equal-tempered scale, the note B4 is at 492.88Hz) and the circuitry necessary to deal with numbers like that and produce tones with such accuracy was prohibitively expensive in the 80s.

When designing chips, there is always a trade-off between the number of components (and therefore the cost), the physical size of the circuit, and the features of the chip. To reduce the complexity, keep the physical size small enough, and keep down the cost of the circuitry, various chip designers took shortcuts in how their chips generate tones and how they deal with frequency information. Most restrict the available frequencies to a small number that requires fewer bits to store, and thus can have a smaller circuit that is less accurate.

Some chips use tricks to compensate for this lack of precision: the VIC-20 gives different voices different octave ranges, the SN76489 loses some bass notes to allow more accurate high notes, and the POKEY lets you join two voices together to increase the precision by a factor of 256.

Square Wave Gallery

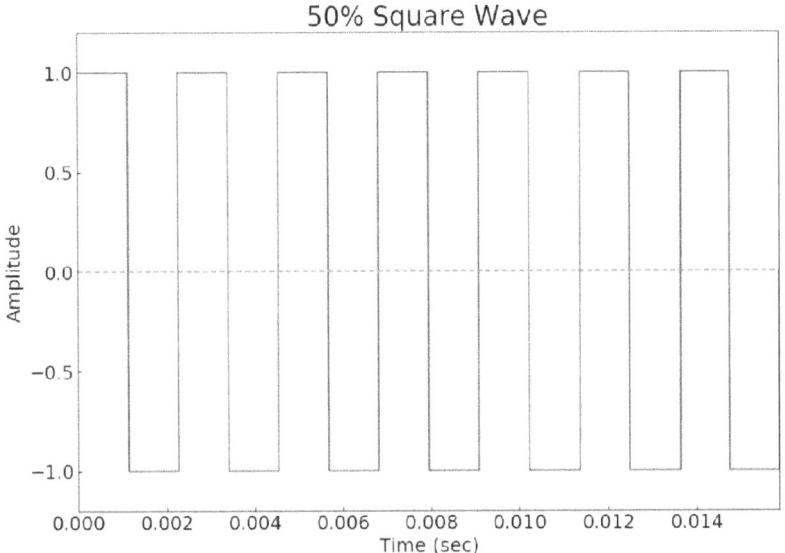

Perfect Square Wave
Square waves are pulse waves with equal amounts of on/off.
They give a slightly metallic sound.
This idealised version has sharp corners and instantaneous transitions from "on" to "off".

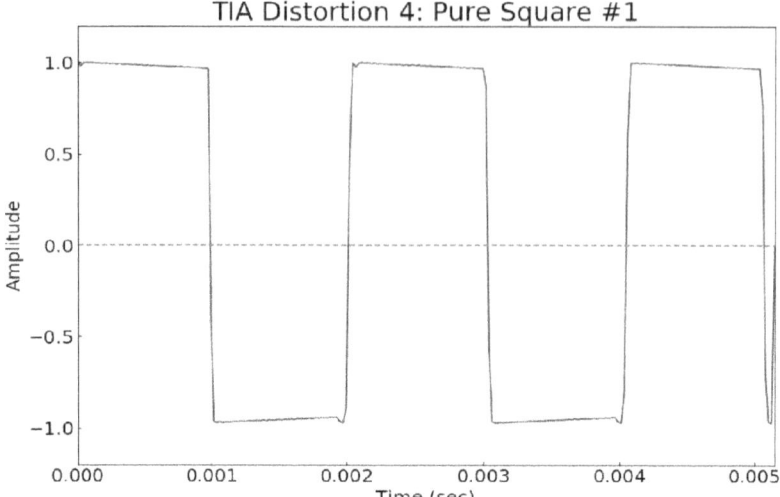

Atari TIA Distortion 4/5
The top and bottom edges of each wave are less stable and more crinkly and the top is slightly sloped.

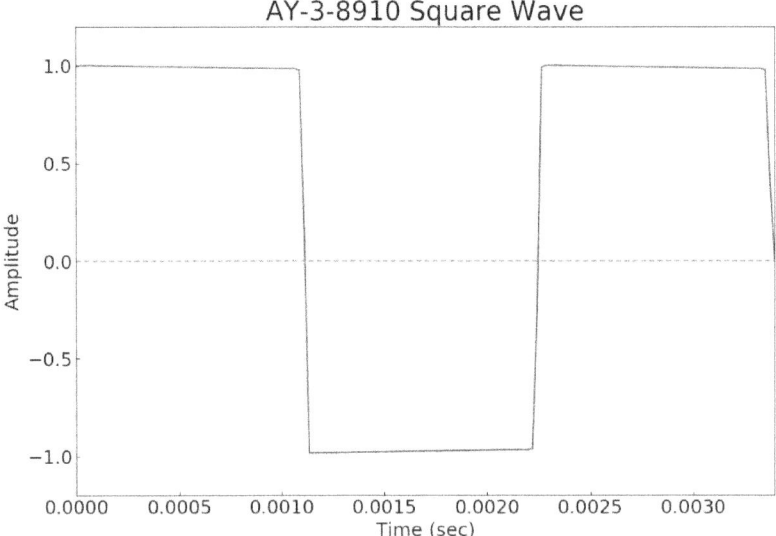

It sounds pure, but it's imperfect, thanks to physics.

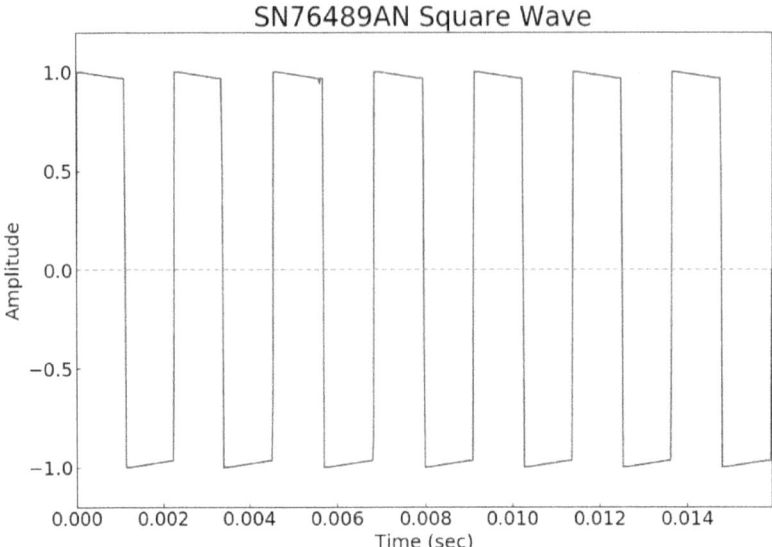

More sloped than the AY-3-891x square wave.

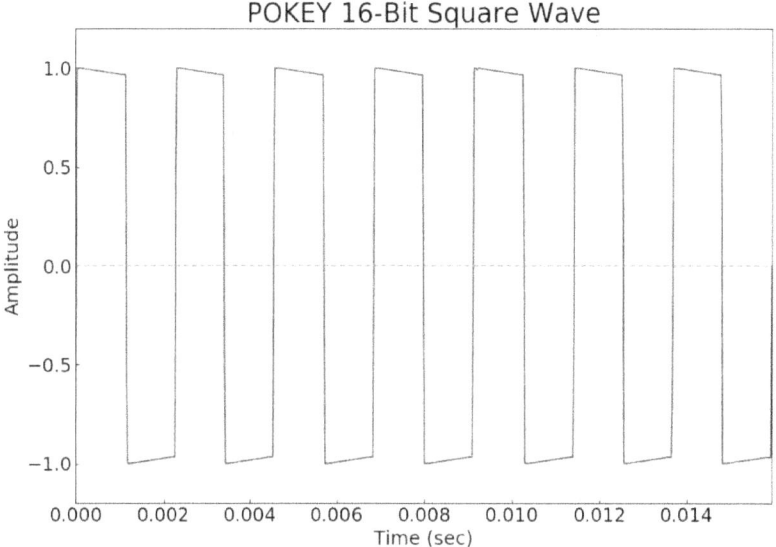

POKEY Almost identical to the SN76489AN wave...
until it's distorted!

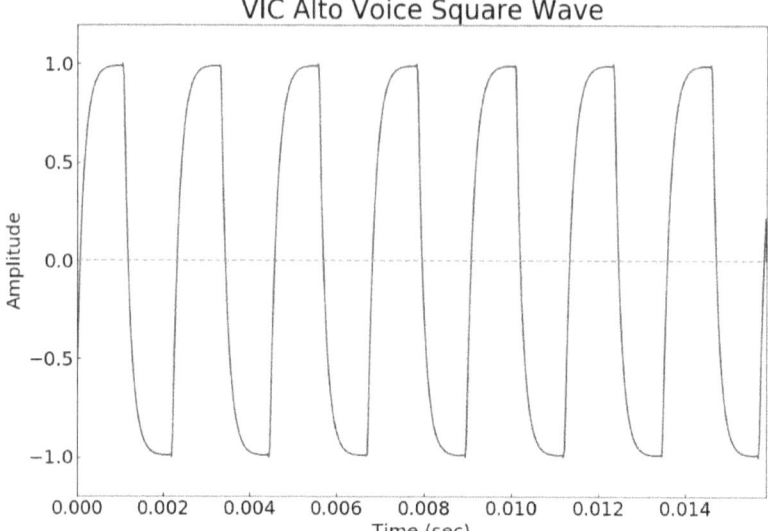

MOS6560 VIC-20 (Alto)
The curved shape explains the VIC's distinctive sound.

(1977) Atari TIA (Television Interface Adaptor)

Iconic two-voice SSG with character and edge. But leave the music to the experts.

As seen in: Atari 2600, Atari 7800.

2-voice SSG with 15 complex waveforms and 5-bit pitch register. Noise channel with 9-bit LFSR.

The Sound of History

Since 1977, the early sound of video games has been traditionally defined by the single-voice melodies and abrasive sounds of Jay Miner's ground-breaking TIA custom chip. This created a stereotype of chipmusic as simple, ear-bashing and "out of tune", a perception that video game musicians have been fighting for decades. However, the raw, apocalyptic sounds blasting out of the television are electrifying, making consoles such as the Intellivision (with a much more capable chip) sound slightly weedy and overly polite.

Mayhem and distortion

This mayhem is output on 2 channels with 15 complex waveforms/distortions, with additional ear damage from a noise channel with a 9-bit LFSR (linear-feedback shift register). An LFSR is often used as a noise generator and how they work is explained at the back of the book!

The TIA on the Atari was originally going to output through two speakers within the case. On some early models, the spaces are there within the case moulds. That idea was scrapped in favour of mono output to the TV, which undoubtedly produced a louder sound. Five titles on the platform, including *Space Invaders*, always output their sound in stereo, and some emulators now support this.

Music is difficult on this chip. The TIA was built for sound effects, and the mechanism for dealing with frequencies/pitches and waveforms is deeply eccentric from a musical point of view. Each waveform is called a "distortion" and there are 15 of them.

TIA Distortions

0 & 11	Inaudible
1	Sawtooth-like, buzzy.
2	Idling tank.
3	Engine noise.
4 & 5	Pure square wave (high).
6 & 10	Slightly buzzy square.
7 & 9	Reedy tones, much brighter. Main *Pitfall* noise.
8	White noise/explosions/lightning/jet engine/spacecraft engine.
12 & 13	Pure square wave (low) that can go much lower in pitch than square waves 4 & 5.
14	Low, slightly buzzy square wave with rumble potential.
15	High buzz which is also ready to rumble.

What this distortion seems to do is to vary the time period and shape of the square waves at different speeds and using different patterns. The Atari 8-bit Home Computer System (HCS) performs the same trick but with fewer variations.

The musical bad news from the TIA is that there are only 32 notes available in total (imagine a piano with only 32 keys and not all the notes in tune).

Also, different notes are audible with each distortion, *and* each distortion behaves differently at high frequencies than low ones.

Decades later, when there were widely available tools, documentation and emulation, amazing feats of engineering such as *TIATracker* appeared, allowing mere mortals to produce passable TIA music.

Frogger VCS - Notes, not croaks!

Dave Lamkins was the lucky person who got to port *Frogger* to the Atari VCS for Parker Bros. In an interview with DigitPress, he also mentioned the TIA's limitations:

> "The Frogger theme music was the second thing I did at PB, right after finishing the development system hardware and software.
>
> The sound chip on the VCS used a programmable frequency divider to generate musical tones. There were only five bits of resolution in that divider.
>
> The two highest notes were an octave apart, and lower notes got closer together. The frequency ratio between the two lowest notes was 31/32. Obviously, this arrangement was not going to give you notes on an even-tempered scale.

> *I had to tinker with the melody a bit so all the notes would be in tune when played on the VCS. If you listen to the arcade game and the VCS game side by side, you'll find that the music is similar, but not identical... the process of adapting the theme music to the VCS was still largely trial-and-error."*

Music... why bother?

The original TIA limitations didn't bother most developers though. Gamers were not expecting advanced music, and the shortage of CPU and memory resources (and the necessity for constant sound effects) meant that full music was impractical. Most 2600 software of the 1982-83 era prior to the "Great Crash" was third-party shovelware. There was no financial reward for raising the bar on music.

> *"Doing music and sound effects didn't take up a lot of my time... there wasn't much video game music that I found at all inspiring. I came of age in the era of acid rock and hard rock music, and cartoony theme music just didn't excite me."*

A notable creative effort, though, was *Gyruss* (also from Parker Bros). Regarded as one of the best soundtracks to grace the chip, it was a magnificent conversion of an arcade machine that had already raised the bar for audio generally with its five AY-3-8910 chips.

Activision took music seriously too.

For the game *Pressure Cooker*, they hired an advertising agency jingle writer to compose the catchiest tune they could, specifying the limited number of notes that were available.

Pitfall II - raising the bar too late!

In one late major release, Activision also raised the bar by removing it entirely in *Pitfall II: Lost Caverns*. This game simply skipped the TIA's limitations by incorporating a custom "Display Processor Chip" into the cartridge, designed by the legendary David Crane. It was originally meant to appear in multiple games, but the "Great Crash" ruled this out. Sonically, this chip featured three square-wave channels and used the TIA for the percussion and sound effects.

In an AtariAge forum post in 2013, the man himself had this to say (note: this gets technical!):

> *"The 2600 had the ability to use the TIA sound chip as a 4-bit D to A (digital to analog converter). One turned OFF the TIA's square wave audio circuit, leaving it in a logic '1' state, and adjusted the 4-bit volume control. Set the volume to 0 and the DC level of the output of the sound chip is zero. Set the volume to $0f (1111) and the DC level output of the sound chip is at maximum. Modulate the audio output between the 16 values of $00 to $0f, and you could generate a waveform. With this method, one could create almost understandable speech — with the consequence of stopping the game and turning off the display since the CPU's entire processing time would be taken up by the process.*
>
> *For the DPC, I designed three independent, free-running clock dividers similar to the two in the TIA itself. One sets a divide counter value and divides the system clock by n to create a square wave in the audible range. Once set, there are three asynchronously running square wave pulse streams being generated which, if you could hear them, would be audible tones.*

So now the big question: how to get those tones from the cartridge and into the TIA chip (and ultimately to the TV's speaker)?

This was accomplished first with combinational logic that simulated an adder circuit. The best way to understand this circuit is to imagine that the first of the bit streams has a value of 6; the second a value of 5; and the third a value of 4. So the first bit stream is outputting 0,0,0,0,0,6,6,6,6,6,0,0,0,0,0,6,6,6,6,6... and so on. Now add the three numbers together.

A1	A2	A3	Value
0	0	0	0
0	0	4	4
0	5	0	5
0	5	4	9
6	0	0	6
6	0	4	10
6	5	0	11
6	5	4	15

The result is a 4-bit number that represents the mixing of three independent tones. It even has the added benefit of providing different relative volumes between the tones. Channel one is slightly louder than the rest, so it is used for melody. The other two channels have the right volume mix for harmony and bass.

Now, how do we get this 4-bit value into the TIA? The 4-bit value is presented to the CPU bus through an address on the DPC chip. In other words, if the code reads a specific address, the value changes each time it is read as the data changes with the music. One could simply set up an infinite loop of 'LOAD ADDRESS, STORE VOLUME' and three-part harmony music will play out of the TV speakers.

But that is not very useful. If all the CPU is doing is reading and writing audio samples, there is no display and no game. So I had to introduce a sampling system. If that 'LOAD/STORE' could be performed at least as often as every scan line of the TV signal the audio would be updated at a sampling rate of 15,750 Hz. This is sufficient to achieve reasonable fidelity.

Using that fact, I inserted that "LOAD/STORE" into every line of the display kernel, every game play subroutine, every line of setup, vertical sync, overscan, etc. I often had to put multiple LOAD/STORES into long calculation routines to make sure that the sample rate was kept up. If not, the music would begin to get fuzzy or even introduce harmonics between the notes and the sample rate. With a little trial and error, I was satisfied with the quality of the musical score in Pitfall II.

I created the DPC chip because the poor old 2600 had lasted years past the lifespan anyone had ever hoped for it. And if we were going to keep making games for the system, it needed a bit of a facelift. Sadly, after Pitfall II, sales of 2600 games finally fell below the point where a game could be made profitably. It would have been fun to see what else we could have done with the technology."

David Crane, August 2013

"P.S. The TIA has two audio channels, and among the built-in 'tones' available is a noise generator. While the first TIA audio channel was busy working as a D to A, I used the second audio channel in noise mode to create a snare drum track. But any time the game needed to produce a sound effect, like Harry's jump, I needed that sound channel. Has anyone ever noticed that the snare drum goes away whenever the game plays a sound effect?"

After the crash

Two products later raised the bar by including synthesised speech: *Open Sesame* and *Quadrun*.

The TIA was also included in the later Atari 7800 console, where it was the only onboard sound generator. Famously, two cartridges for it included POKEY chips: *Ballblazer* and *Commando*.

Both the ColecoVision and Intellivision had add-ons that allowed them to play VCS cartridges. Both add-ons contained a version of the TIA. Mattel used a new technology called "chip-on-board" (COB) to make sure Atari couldn't prove the TIA design had been stolen!

The chip did see one use in the arcade, as the 2600 title *Video Olympics* was released in the arcades by Atari in 1978 as *Tournament Table*.

Unlicensed clones also saw retail release in the 1990s (and beyond) initially through the "TV Boy II" and "Super TV Boy", with many original 2600 releases included. These were often renamed and/or hacked to remove copyright notices and other assets. This seems a very naïve way of trying to avoid legal action!

TIA Distortion Gallery

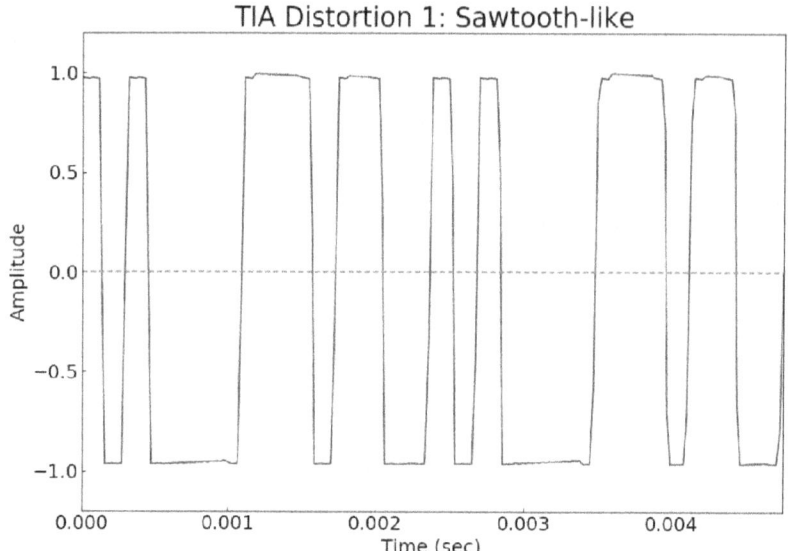

Sawtooth-like, buzzy waveform.
The occasional decreased-width square waves
produce a chirp that sounds like a sawtooth wave.

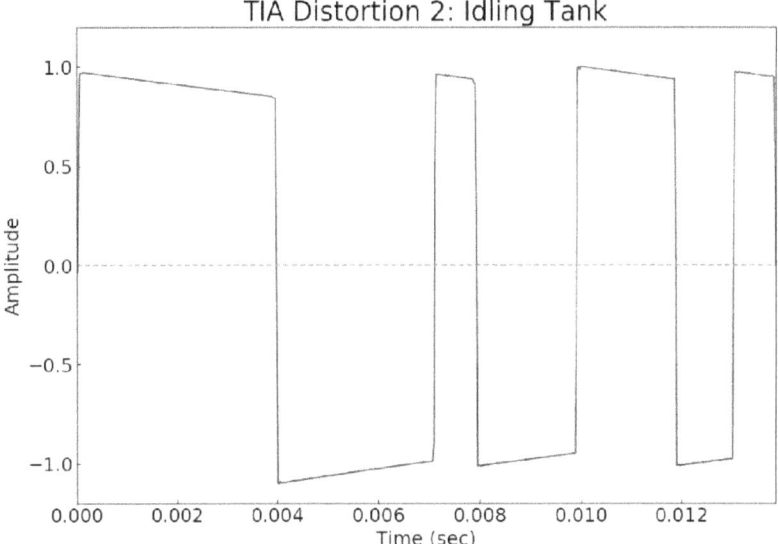

Like an idling tank.
You can bet they used this in the game Combat!

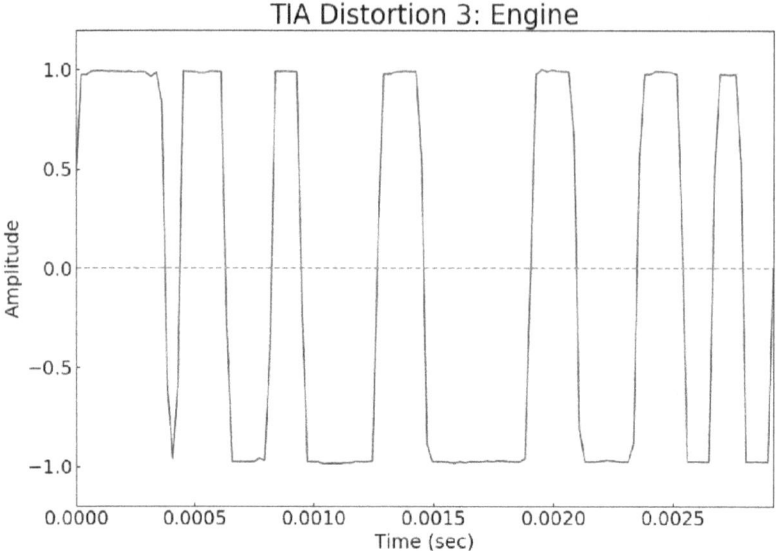

Engine noise. Many racers rely on this!
Square waves are closer together and change width quicker for a rougher sound than the idling tank.

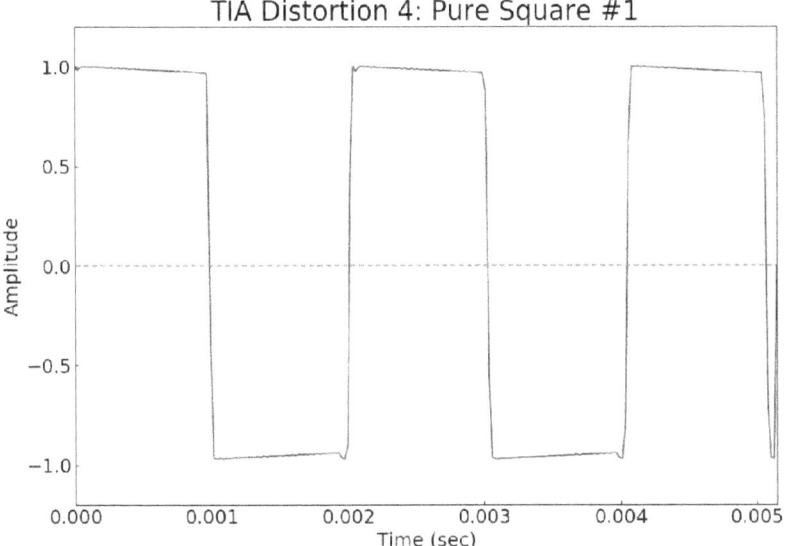

As pure a square wave as the TIA can produce.
This has a higher range than the other square wave distortion.
The evenness of the timing indicates an unchanging pitch.

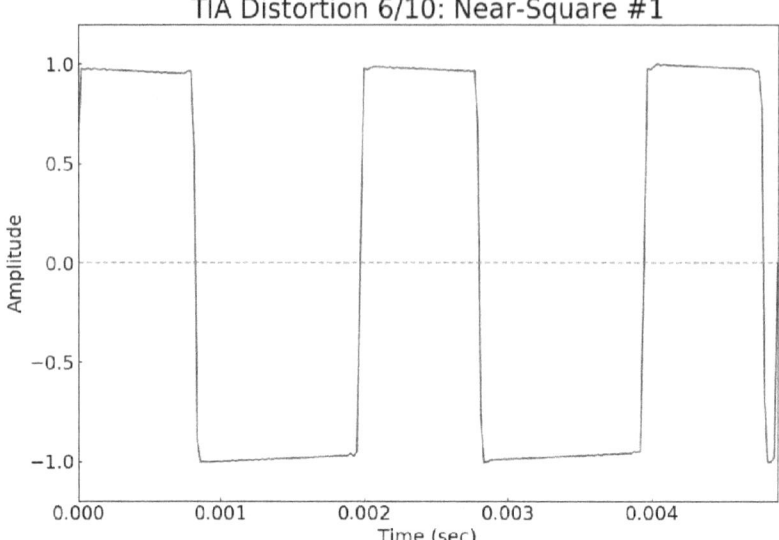

A slightly buzzy square
The spacing has become less even between the top and bottom peaks.

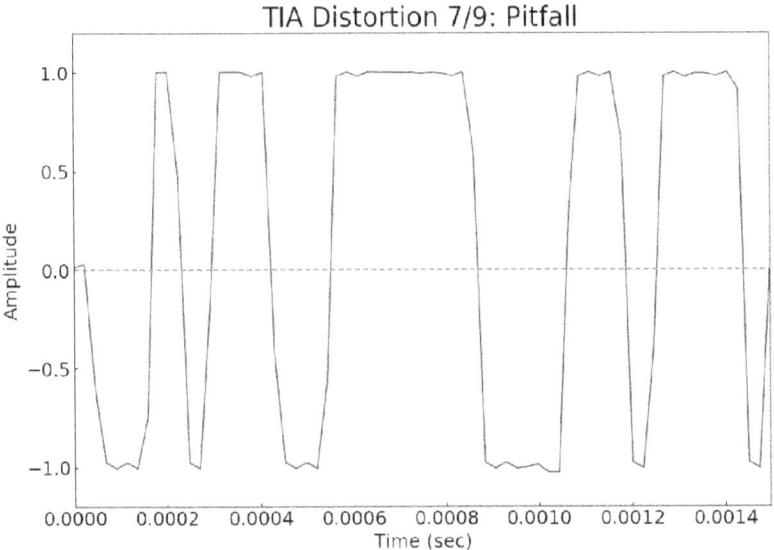

Reedy tones, brighter than the usual square wave.
Apparently, it's one of the main noises in Activision's Pitfall.

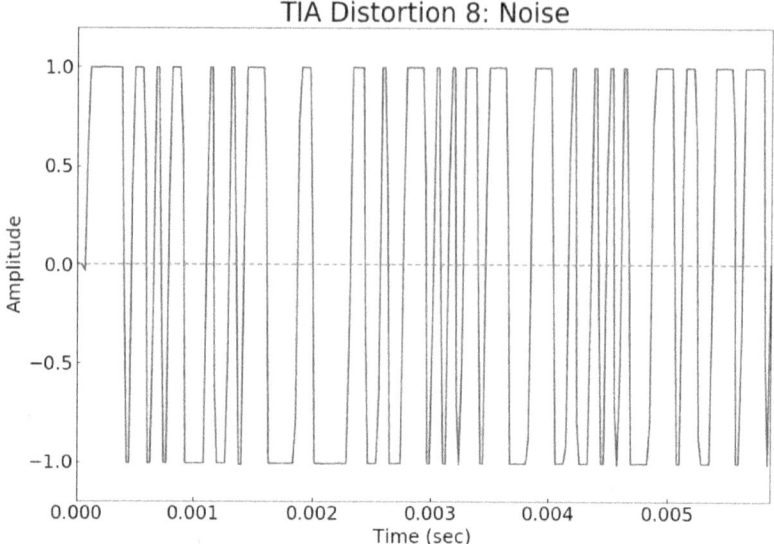

This is the TIA's "White Noise" setting.
As we've seen from the spectrogram, this noise appears pretty solid.

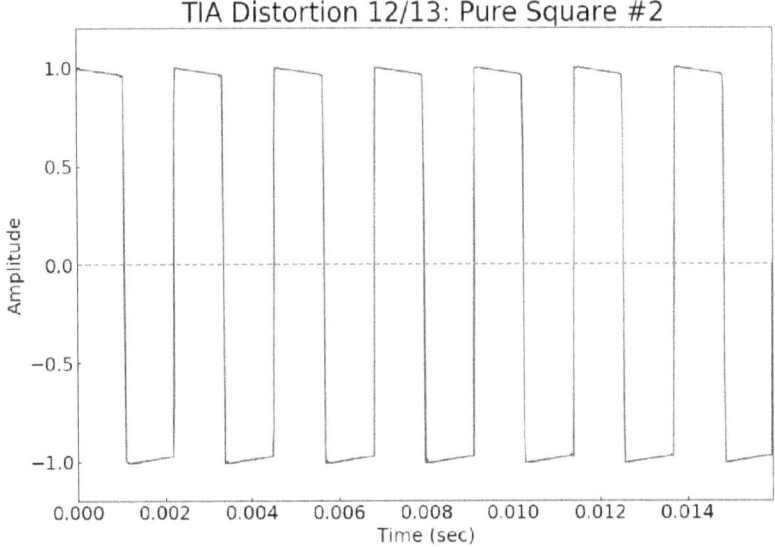

A pure square wave that can go to a lower pitch than distortion 4 & 5. This would be the bassline distortion!

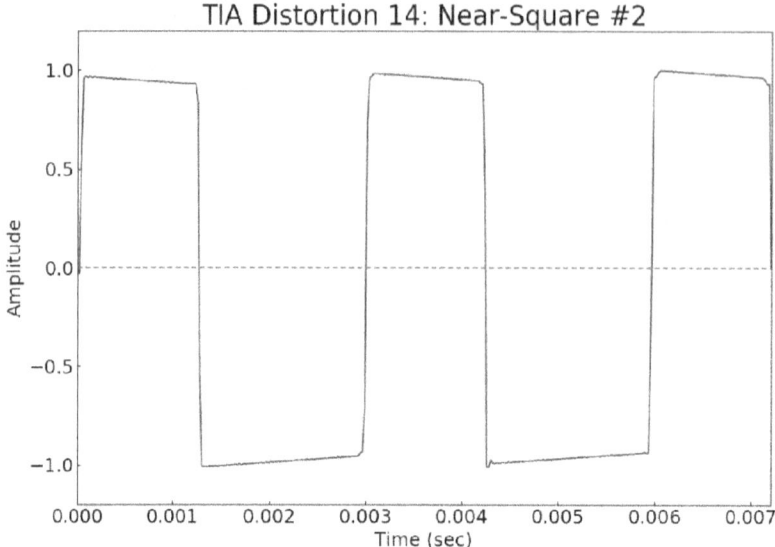

Bottom troughs are wider than top peaks,
essentially changing the pulse-width.

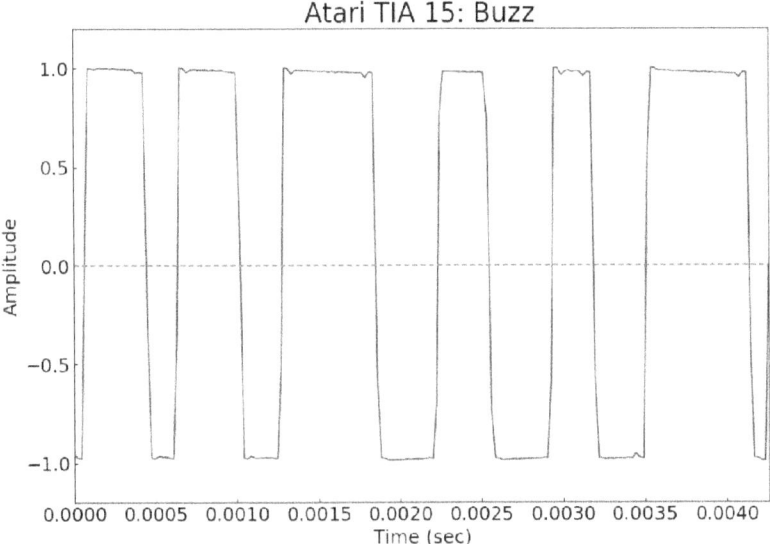

I'm not sure if the TIA needed more buzzing sounds, but here's another one anyway.

(1978) Intel 8244 (NTSC)/8245 (PAL)

Pioneering single-voice square wave generator; accidentally, the first wavetable chip.

Single-voice, 3-byte, 2-frequency wavetable playback from a 24-bit shift register, with noise generator. It's also a video chip! Mono.

An Odyssey to Innovation

> *The sound generators worked by placing appropriate bit patterns into recirculating shift registers. This gave maximum flexibility while minimizing the computer program's work. Sounds are an important ingredient to video game players."*
> **IEEE Annals of the History of Computing, July–September 2009**

The 8244 was exclusively designed for Magnavox, and, as we've seen with the Atari TIA, it often makes sense to create a single chip that combines audio and video capabilities when designing custom silicon. In this case, designer Peter Salmon, alongside up-and-coming chip manufacturer Intel, had a brief from Stanley Mazor to keep the price of the chip under $20. Every decision therefore prioritised that goal, leading to some complex and novel engineering.

> "The 8244 video-game chip had 40,000 transistors. The random logic" was hand drawn using color pencils."
> **Stanley Mazor**

The Odyssey 2 - making music difficult

The result of this engineering was the Magnavox Odyssey 2, one of the "big three" cartridge console systems in the US in the late '70s/early '80s, behind the Atari VCS and the Mattel Intellivision.

Craig Kubey described this console in his iconic 1982 book "The Winners' Book of Video Games" as "… an inferior product", pointing out that the games are neither as famous and exciting as on the Atari VCS, nor as "realistic" as the Mattel Intellivision. A keyboard partially makes up for it.

While the audio portion of this chip is innovatively engineered, it isn't built for music: there's no built-in mechanism for selecting pitches.

It uses a very early form of wavetable synthesis, sending a user-defined sequence of bits (on/off) quickly enough to an output (TV or speaker) to generate an audible sound. While in later wavetable synthesis this could be done at many different speeds to generate musical tones, this chip gave you two speeds: high or low. The output could be looped, and you could use a pseudo-random noise waveform instead of the tone generator.

The system has a small musical jingle of ascending tones as a start-up tone, both to delight the consumer and bring relief to any engineer trying to fix it. Even a simple jingle like that must be laboriously engineered, given the chip's inherent musical limitations.

Since memory on the Odyssey is extremely tight and most programmers weren't also sound engineers, they usually took advantage of the eight preset sounds available in ROM: "beep/error", "explode", "alarm", "select", "keyclick", "buzz", "select 2", and "shoot". The latter sound is reminiscent of Oric BASIC, which had "Zap", "Ping", and even "Explode".

The 1980 cartridge *Videopac 31 (Musician)* contains the programming to generate actual musical tones and provide a metronome, and it also included a home-organ-like keyboard overlay.

The chip was also deployed in the Videopac G7000 in Europe (C52 in France) and made a re-appearance in 1983 on the Odyssey 3 Command Centre (Videopac+ G7400 in Europe, Jopac JO7400 in France), by which time it was competing with vastly superior technology.

(1978) Texas Instruments SN76477

Quirky single-voice analog SFX chip responsible for some of early gaming's most distinctive and iconic sounds.

As seen in: Space Invaders (UFO Sound), Luxor ABC80, Hector/Interact/Victor Lambda.

Single-voice mono analog SSG with square/noise waveforms, a low-pass filter, and an attack/decay envelope (no sustain/release settings).

Sound Invader

One of the first off-the-shelf audio chips, this chip was in use even before it was released to the world, embedded in the iconic arcade game *Space Invaders*.

The development of *Space Invaders* was a key factor behind the chip's origins as a collaboration between Taito and Texas Instruments, though the chip was only used for one sound-effect (the UFO)! The Space Invaders PCB treated each sound as a discrete analog circuit (with its own volume control), including the SN76477.

Analog awkwardness

This chip is an awkward hybrid analog/digital chip, since it's primarily configured by resistors, capacitors, and voltages.

This made it more difficult to control with a microprocessor than later chips, such as the ubiquitous-yet-anonymous SN76489, a four-voice, digital tone-fest, released to compete with the ubiquitous 1978 AY-3-891x family.

Although this single-voice chip is theoretically capable of producing noise and square-wave at the same time, the mixer partially defeats this by essentially making them fight each other. However, it does occasionally yield useful sonic results.

A patent filing reveals a low-pass filter was added for:

> "… smoothing out the essentially square wave-like waveforms produced by the digital [logic]… This tends to eliminate the common objection to synthesized sounds that they sound electronic".

Of course, the chip sounds electronic anyway! A simple hardware envelope was also added to change the volume of the sound over time, adjusting how quickly it "attacked" and how quickly it "decayed".

Taming the beast

The chip itself was not marketed as a chip for making music. While discrete tones can be coaxed from it, it's not straightforward because of the analog nature of the chip. There's a direct linear relationship between voltage and pitch that is hard to control, especially to produce higher pitched notes without sounding out-of-tune.

However, as with most technology, it was later abused to do clever things it was not designed to do. For instance, an SN-VOICE PCB was created by a hobbyist named Thomas Henry that could provide accurate musical notes for a 5-octave range. Scene demos released for the ABC80 home micro prove what the chip can do in skilled hands.

Numerous variants of the chip were released, including a narrow version (SN76477N) and a version with its own speaker (SN76488), which could be easily integrated into toys. Fans of the chip have also designed it into other analog electronic devices such as standalone synths.

Arcade adventures

With higher per-unit budgets, there are numerous cases of arcade machines employing multiple sound/music chips, either different chips for different types of sound or multiple chips of the same type. A few of the later arcade games such as *Stratovox* (the first arcade game featuring speech) used it for wild analog sounds, but also added in an AY-3-8910 for music. *Sasuke vs. Commander* contains three SN76477s driven by a 6502 instead of a Z80, an unusual move.

Apart from *Stratovox*, some pinball machines, and *Space Invaders* (which spawned a ufo-load of bootlegs such as *Space Wipeout, Super Invasion, Alien Invasion* and *Cosmic Monsters*), this chip also served in coin-op games such as *Space Launcher* and *Sheriff* (Nintendo), *Space War, Indian Battle* and *Crazy Balloon* (Taito), and *Laser Battle/Lazarian* (Zaccaria).

Hiding in the home

Despite the difficulty in controlling the SN76477 with a microprocessor, there were some home computers that included it, such as the Swedish Luxor ABC80.

Another was the short-lived 8080A-based Interact Family Computer, released in 1979 shortly before its parent company Interact went bankrupt. When it was sold in France, you had to buy a US television as well! Later modifications to make it work with French TVs made this the first home computer with a SCART socket.

After both Interact and Lambda had gone bankrupt, Toulouse-based Micronique acquired it and reworked it as a Z80A machine.

It was relaunched in 1981 as the Victor/Hector, followed in 1983 by the Hector 2HR ("High Resolution"). Hector MX40c and Hector MX80c models followed in 1985.

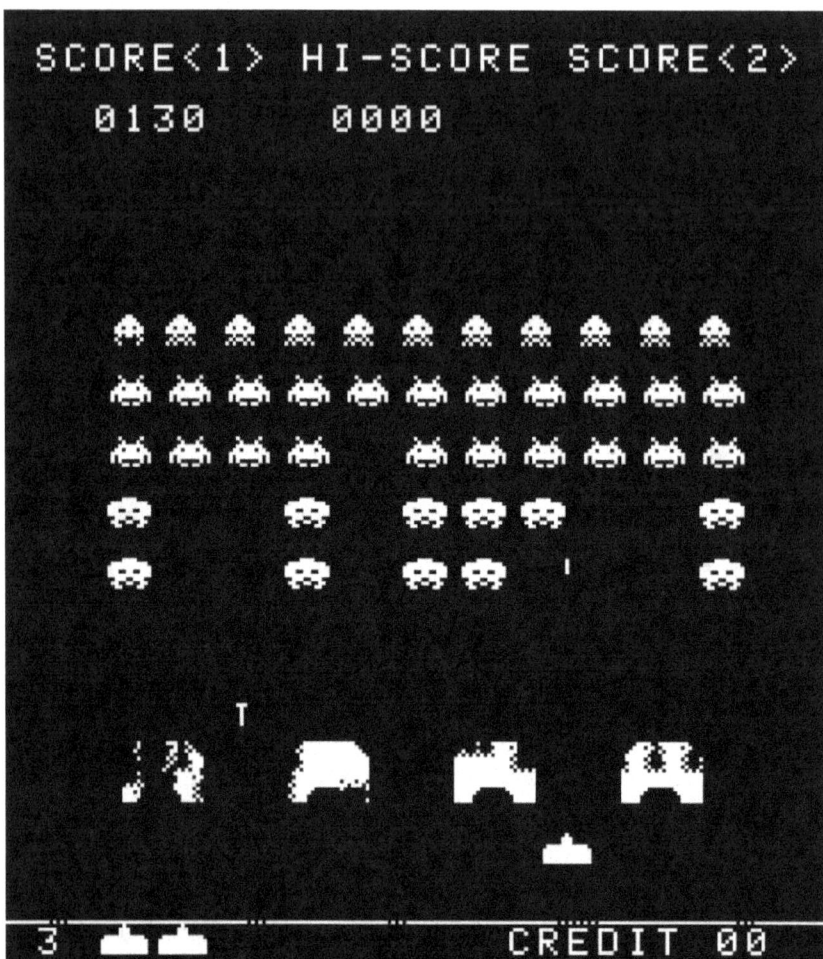

(1978) Bally-Midway 0066-117XX

Overlooked but brilliant 4-voice SSG for the time.

As seen in: Gorf, ABA-1000 (also known as Bally Astrocade & Bally Professional Arcade).

SSG with 3 pulse/square channels and 1 noise channel. Mono output, except in the arcade where the noise channel could be routed separately.

An Arcade in Your Home

In the early days of consoles and home microcomputers, an "arcade in your home" was the biggest promise made by console manufacturers, and it was not an easy one to keep, given the gulf between arcade hardware and the hardware fitted in home consoles. No company in the early days tried harder than the Midway side of Bally-Midway, who, in 1976, contracted Dave Nutting Associates to design custom hardware for them, both for arcade and home use.

Nutting delivered a Z80-based full-colour graphics system with a custom sound chip that was first deployed in (possibly) the first full-colour arcade game *Sea Wolf II* (1978).

Later, it was used in some of Midway's classic arcade games of the era, including *Gorf*, *Wizard of Wor*, *Space Zap* and *The Adventures of Robby Roto*! The speech in *Gorf* and *Wizard of Wor* came from a separate chip. The last Astrocade board deployed to the arcades was the "not approved by Namco" quiz game *Professor Pac-Man* in 1983.

With a top video resolution of 320×204 for the graphics and a three-voice sound-chip with built-in hardware vibrato, the board was indeed worthy of the arcades, but you needed expensive RAM to use it at these specifications.

Three voices and vibrato too

Dave Nutting's "Music Processor" chip was sophisticated and powerful: three square wave generators and a separate noise channel that could be mixed in with the other square waves, if necessary. The addition of a hardware vibrato is an indication that music generation was on Dave's mind. Hardware vibrato was also incredibly useful for wavy and imaginative sound effects and didn't use up valuable CPU time.

The chip was mostly used for sound-effects and small jingles. The chip had 256 possible frequencies/notes available to it (as opposed to the TIA's even-more-restrictive 32 notes). This relatively small number of available frequencies means that tuned notes can get increasingly out of tune as they get higher. True accuracy at high octaves requires either more precision or workarounds.

Midway's chip uses a master oscillator applied to all three voices, like a chair that the other voices stand on to be able to reach high places more accurately (at the expense of bass coverage).

Bally Home Library Computer Professional Arcade Computer System Astrocade… or something!

The clunkily-named "Bally Home Library Computer", announced in 1977 but launched in late 1978, was the only computer that included this chip. It was later renamed "Bally Professional Arcade", then later re-released as "Bally Computer System" then finally renamed "Bally Astrocade".

Its 1978 price was a chunky-but-not-unreasonable $299, and it was a full-spec home computer: QWERTY keyboard, programmable in BASIC, controllers and other peripherals.

Unfortunately, the promised arcade-at-home experience was only partially delivered. The screen could only display 160×102 resolution thanks to the 2 MHz RAM speed, and only 160x88 in BASIC.

How on earth do you fit BASIC code into a machine with only 4 KB of RAM, some of which is for the graphics? In this case, you reduce the width of the screen in BASIC and hide the code in the display memory at the sides of the screen! Ingenious.

For a while, this was the cheapest home computer in the world, but it was a commercial disappointment. It was marketed poorly and the market was lukewarm to start with, so it sold only a few thousand units.

Ahead of the market

In fact, this micro wasn't alone in 1978. The VCS was also regarded as a sales disappointment, only catching fire when Space Invaders was released in 1979. It's reasonable to think that both systems were ahead of their time, especially the Bally, which had to sell the idea of a home computer as well as selling itself.

(1978) Signetics 2636 PVI

A vaguely musical single-voice beeper.

As seen in: The 1292 Advanced Programmable Video System, Hunchback, Lazarian, Superbike.

Single voice SSG. Mono.

Have you heard of the "Interton Video Computer 4000"? Or the "1292 Advanced Programmable Video System"? Would you know their dulcet square-wavey tones if you heard them?

Disappointing stocking fillers

In 1978, if you couldn't afford an Atari VCS, there wasn't a shortage of off-the-shelf cartridge-based consoles to choose from. Names like Grandstand or Grundig were cleaning up at the cheap end of the market, selling consoles and Pong machines to parents and grannies who had no idea which console their child wanted for Christmas.

On the other hand, all video games were exciting for children at the time!

In the UK, this chip was in the "Rowtron Television Computer System" (aka "Teleng Television Computer System"), one of which was owned by composer Mike Clarke:

> "We played it a lot. It had simplistic, yet compelling, games."

There were numerous members of this family of chips, all with different software compatibility. They all originated from the chipset in the original 1978 version from Europe's Audiosonic, based around the Signetics 2650 CPU (released in 1975!) and its companion graphics and sound chip, the Signetics 2636 PVI.

It's square. It waves. It beeps.

There isn't a lot to say about the sound capabilities of this chip. You have a square-wave beeper with an 8-bit frequency register, and that's it. The 2636 chip doesn't have a DAC to convert that to audio. Instead, it's passed back to the CPU for mixing into the UHF signal destined for the TV.

Hunchback at the office

The most notable game built around this chip (deployed in Century's upgraded version of this system, CVS, aka Convertible Video System) was the arcade version of *Hunchback*. It also powered coin-op clones such as *Galaxia*, *The Invaders* and *Logger* (a *Donkey Kong* clone, not *Frogger*!), which makes you wonder where all the IP lawyers were in the early '80s.

CVS cheated a little to make its games arcade-quality, using a triple 2650/2636 combo: one for the game and graphics, one for the sound, and in some deployments, one to control a Texas Instruments speech chip.

Because *Hunchback* was later converted to run on other arcade boards (such as *Galaxian*, *Donkey Kong* and *Scramble*), not all versions had samples and speech. The top-of-the-range Century Electronics cabinets featured speech and sounds, such as the swish of the swinging rope, but that was produced by a DAC sample rather than the under-powered 2636 PVI.

Rowtron Games
1. Sportsworld
2. Combat
3. Horse Racing
4. Maze
5. Maths Two
6. Four in a Row
7. Mastermind
8. Air-Sea Battles
9. Boxing
10. Blackjack
11. Sporting Shotgun
12. Motor Race
13. Maths One
14. Circus
15. Galactic Space Battles
16. Reversi
17. Video Pinball
18. Flag Capture/Memory Match
19. Face the Music
20. Golf
21. Draughts
22. Alien Invasion
23. Home Programmer
24. Cowboy
25. Chess
26. Monster Muncher

(1978) General Instrument AY-3-891x

The Coca-Cola® of sound chips.

As seen in: Intellivision, MSX1, Spectrum 128, Gyruss coin-op and many more!

SSG with three square-wave voices. Noise and ADSR envelope are shared between the voices.

Full subsonic-to-supersonic note range. Noise is from a 17-bit linear feedback shift register. Individual voice outputs have their own pinouts, and can be mixed to mono or stereo, depending on implementation. Most implementations mix the sound to mono.

We're #1! We're #1!

If your metric for #1 sound chip is the number of deployments, the unassuming AY-3-891x family is the undisputed king of chips, included in over 2,400 arcade titles (more if you include the rebadged YM2149F version and chips such as the YM2203, which bolted on FM voices).

It dominated the early '80s, peaking in 1984, before use of the chip declined in the face of competition from Yamaha themselves and their new FM technology, though these more advanced chips were often deployed with an AY for additional sound effects or even just for the timers! Some Yamaha FM and PCM chips even had an AY core at their hearts, based on the compatible YM2149F that was best known as the sound chip in the Atari ST.

Some composers nicknamed this chip "The Doorbell Chime" due to its square wave generators giving it a familiar sound quality, though this nickname was more applicable to later FM chips that sounded much more bell-like and clangy.

For future release: the Gimini Deluxe

The AY-3-891x family became the most ubiquitous sound chips ever created, but it's unlikely General Instrument had any idea that would happen since this iconic chip was first marketed as an optional add-on!

The GI Microelectronics data catalogue from 1978 — the first one after the chip was created — had a section for "Programmable TV Games", encouraging companies to launch their own consoles based on the reference design.

The star of the page was the Gimini Deluxe "8900" Programmable Game System, built around their "16-bit" CPU CP1610. Optional extras available to OEMs included "Cartridge ROM", which was a 2048 x 10 ROM containing additional game programs.

The TV interface ("for future release") was called "AY-3-8900", the optional colour processor (also "for future release") was "AY-3-8915", the optional cassette tape interface (yet again, "for future release") was "AY-3-8920" and the optional sound generator (which was actually shipping!) was "AY-3-8910" ("provides programmed generation of complex sound effects").

Guess how many customers bought the "Gimini Deluxe 8900"? Only one really matters: Mattel Electronics, who used it as the base hardware for their Intellivision home console, with AY-3-8914 on board.

Mattel had the resources to compete with Atari on their own turf, though with more of a focus on realistic sports games, and as a result the Intellivision was the #2 console in the late '70s/early '80s. It wouldn't be surprising if its success was a great advert for the AY-3-891x, giving the General Instrument marketing department a big win.

In home microcomputers, it took until 1983 to achieve critical mass when it was made part of Microsoft and ASCII's "MSX" standard specification for computer manufacturers. However, it was present in some micros before that, including in the dual-AY-3-8910 "Elektor TV Games Computer", the "Color Genie", and the "Jupiter Ace".

Examples of microcomputers and consoles containing AY-3-891x chips

Year	Model	Chip
1979	Intellivision	AY-3-8914
1981	Elektor TV Games Computer (Expanded Version)	AY-3-8910 x 2
1981	NEC PC-6001	AY-3-8912
1982	Votrax Personal Speech System	AY-3-8912
1982	Color Genie	AY-3-8910
1982	Vectrex	AY-3-8912
1982	Intellivision II	AY-3-8914A
1983	MSX1 Standard definition	AY-3-8910
1983	Oric 1	AY-3-8910
1983	Timex Sinclair 2068 (US) - the first marriage of Sinclair Technology and AY!	AY-3-8912

1983	Aquarius Mini-Expander	AY-3-8914
1983	Spectravideo SVI-318	AY-3-8912
1983	HT-1080Z Series I	AY-3-8912
1983	HT-1080Z Series II	AY-3-8912
1984	Amstrad CPC Family	AY-3-8912
1984	Oric Atmos	AY-3-8912
1984	Tiki Data 100 (Norway)	AY-3-8912
1984	Sharp X1 Turbo (CZ-850C)	AY-3-8912
1985	SNK HAL21	AY-3-8912
1985	HT-1080Z/64	AY-3-8912
1986	Oric Telestrat	AY-3-8912
1986	ZX Spectrum 128	AY-3-8912
1990	Amstrad GX4000	AY-3-8912

Setting the standards... then defeating them

Being the first sound chip capable of producing the entire audible frequency range, this chip is almost a reference in the realm of chipmusic, and it set the standards (and the sounds) in SSGs.

The focus on sound effects in the original marketing is a big clue that music was not top priority when designing the chip.

At that time, there was no credible intersection between the worlds of music and video games. Electronic musicians of the time (who were themselves the objects of prejudice from established musicians) were enjoying full analog synthesizer programming (with knobs!). The world of digital sound offered nothing but pain and ridicule, especially when getting the most out of the technology required you to learn assembly language. To the hobbyist of the time, this chip offered beeps, hiss, and not much else, even if there was a lot of unexploited power under the hood.

It could also be abused. For example, programmers would configure the hardware envelope for cycle times above 20 kHz to produce sawtooth or pulse-wave-like bass sounds. Indeed, later musicians/programmers often abused, disregarded, or misused hardware features or implemented highly controllable software features run by their "driver" (the code playing the music "live").

Musicians such as cross-platform specialist David Whittaker and Ocean's Jonathan Dunn even produced some work that people preferred on the AY chip to the Commodore 64's SID, in particular *Feud* (Whittaker) and *Robocop* (Dunn). The Gameboy version of *Robocop,* played through a different but equally square-wavey chip, was later repurposed for a fondly remembered "Ariston" UK TV commercial in which a short loop of this tune played while repetitive activities occurred involving the use of white goods.

The Follin brothers Tim and Geoff were also highly regarded as AY chip wizards.

Not just for Mattel

This chip spawned a family of successors. Repackaged versions and functionally identical clones are still being produced!

General Instrument variants include the AY-3-8910 (40-pin), AY-3-8912 (28-pin), and the AY-3-8913 (24-pin). This was used on the Apple Mockingbird Sound board which was supported by numerous Apple II games including *Ultima V*, *Adventure Construction Kit*, *Skyfox* and *Zaxxon*). In the Intellivision and Aquarius, Mattel used GI's AY-3-8914.

MSX - the next big thing

The AY-3-8910 was part of the Microsoft/ASCII MSX spec for home computers and most vendors used GI's chip. However, compatible silicon was sometimes deployed, such as a Toshiba chip (the T7766A).

In the mid-'80s, the MSX market for Western software houses was big enough to sell to (especially Spain) but not big enough to lavish resources on. The shared CPU (and later, sound chip) between the ZX Spectrum and the MSX ensured that there were plenty of lazy ports.

> *"Developing a game for a preferred platform and porting it was a common practice of European, Japanese and USA developers. The MSX system did not have a very powerful graphics chip compared to the Famicom PPU or C64 VIC-II. It had no hardware scroll, only four single-color sprites per line and slow pixel drawing.*
>
> *As a result, they coded conversions with a kind of Spectrum screen emulation on MSX, with varying results. 99% of European releases were on tape, and most were*

> Spectrum ports that were much uglier than they should have been.
>
> They were also slow because rewriting the game to optimise it with hardware sprites and colours was too expensive. Some Japanese companies did the same, converting PC88 games to MSX2, such as Elf Co., Snatcher and Glodia.
>
> These used only eight colours out of 16 and were also very slow."
>
> **Toni Gálvez, Graphic designer, Dinamic Spain**

However, MSX users often lost out even when the machines started to share a common sound chip, as Toni explains:

> Even though there was an AY/YM chip, most games were converted from the Spectrum 48k (with beeper sound), which is pathetic for an MSX computer. Also, because the AY chip runs at a different speed on the MSX, tunes from the Spectrum 128 needed to be adapted or rebuilt."

While Toni is right about the different clock speeds, this proved no obstacle to musicians such as David Whittaker and Tim Follin who ported music data as-is from the ZX Spectrum 128 to the Atari ST, resulting in the Atari ST version sounding more stressed, thanks to its marginally faster clock speed.

Going Japanese

Later Yamaha variants of the chip include: YM2149F (40-pin, double resolution), YM3439 (CMOS, 40-pin), YMZ294 (18-pin) and YMZ284 (16-pin). The Winbond WF19054, JFC 95101 and the File KC89C72 have the same pinout as the AY-3-8910 and are also 100% software compatible. They're still in production and used in many slot machines.

The AY-3-8914 has the same 40-pin package and pinout as the AY-3-8910, but minor control register changes. It was used only in Mattel's Intellivision console and their Aquarius computer.

Enhanced

The Microchip AY8930 is an enhanced but mostly-backwards-compatible version of the AY-3-8910. It adds major features such as separate envelopes for the three channels (as opposed to one shared envelope), variable duty-cycles (so pulse-width modulation is possible), more bits of precision for note frequency, volume, and envelope frequency, and a more configurable noise generator.

It was used on the Covox Sound Master sound card for the IBM-PC and the Darky ePSG Stereo Card by Supersoniqs.

Speakers, beepers, and squeakers

A substantial amount of early tech was shipped with a low-quality beeper or buzzer inside the case.

It's pushing the definition to regard a speaker as a sound chip, but it would be unfair not to mention some of the iconic (and less-than-iconic) machines that had a humble beeper-in-a-box.

Computers that shipped with nothing except an internal piezoelectric beeper included the Commodore PET, the Apple II and the ZX Spectrum, as well as less successful machines such as the Camputers Lynx and Jupiter Ace. A large collection of business machines also featured beepers, with IBM PC compatibles being the most prominent. The original PET 2001 model had no sound output, so hobbyists added an output to an RCA jack through pin CB2 on the user port. Commodore adopted the idea on later PET models.

PC speaker support in games was a standard feature until the early 1990s, causing a headache for composers such as EA's Rob Hubbard, who needed to scale their sophisticated scores down to one voice while not losing their essence.

While reviewers and gamers were relatively forgiving of simple beeper sound on the Apple II or PC, developers on the original ZX Spectrum needed to make their games sound impressive even through the speaker. It was a popular gaming platform and quality of sound/music had become a significant factor in the all-important magazine reviews.

A particularly iconic use of the speaker is Manic Miner's continuous "Hall of the Mountain King" that inherits a lot of its charm from resonance created by the ZX Spectrum's case. Its cacophonous, wonky rendition of "On the Blue Danube" was also notable.

Various composers also produced single- and multi-channel music for the base-level Spectrum using the beeper.

The Follins pushed multi-channel-through-single-beeper output further than most through complex composition and sonic ambition. In pieces like *Chronos*, though, the music sounds more coherent in emulation than when played through a beeper-in-a-box.

Tim Follin's multi-channel Spectrum epics were achieved by switching the speaker state based on the highs and lows of each virtual channel's pulse wave at any point in time. Five different registers were used to keep track of the output, a masterpiece of coding.

Of course, any computer with a speaker was fair game for speaker-sample abuse. But what about computers with no sound at all?

No beeper, no problem!

> *"One interesting bit of trivia about the TRS-80 Model 1 was that it was supposed to be silent. Early gamers though wanted the blips and beeps… soon programmers realised you could push sound effects out of the cassette port.*
>
> *All a user needed to do then was to hook the cassette OUT port (normally used for recording programs on audio tape) up to a speaker and there you go!*
>
> *Almost all later TRS-80 Model 1 games contained sound. Some sound routines got quite sophisticated although, in the absence of a dedicated sound chip, their use took a lot of processor time."*
> **Terry Stewart, TRS-80 fan**

Owners of other quiet machines such as the Commodore PET and the ZX81 also cottoned on quite quickly, and for a while every weekly magazine in the UK seemed to have a ZX81 synthesizer listing to type in.

Martin Galway's beeper disaster

> "I sat down in Joffa's office to do Cobra first thing in the morning, with the plan to go back to whatever C64 title I was working on afterwards. He showed me how to put the notes in on the Tatung Einstein dev system, and I started typing away, composing as I was going.
>
> Ocean was continually adding staff and it was not uncommon for electricians to be adding power sockets and stuff.
>
> This was going on out in the corridor while I worked, and after about 2 or 3 hours of solid work and a really great tune being virtually finished... the entire floor had a 10-second power-cut due to somebody not communicating properly.
>
> I realised then that I had not hit "SAVE" the entire morning and the tune was gone! We sorta laughed and went to lunch, and when I came back in to re-do it... I had mostly forgotten it.
>
> So I started again with a new tune and "Cobra" got the tune it shipped with, which got used on Arkanoid after that."
> **Martin Galway**

How to play back samples on something that can't play samples

by Mike Clarke, author of InSIDious VST

Digital samples are played by storing a set of numbers that represent the waveform of a sound and sending them in sequence through a digital-to-analog converter circuit (DAC). This circuit converts each number to its corresponding voltage level, and that voltage directly determines how far out or in your speaker cone is at that point in time. The motion of the speaker due to this changing voltage compresses or expands the air next to it, creating a sound pressure wave that eventually reaches your ears.

The level of detail that you have for the set of numbers determines how good the sound quality is. If you have lots of numbers and can run through them very quickly, you can play high frequency sounds. The speed that you run through them is called the sample rate. The cleanliness of that sound is determined by how wide the range of numbers is. If you can only use a small range, the signal will have more noise. The range of values you can use is determined by the bit depth — more bits allow a wider range.

Compact Discs use 16-bit numbers, ranging from -32768 to +32767, played at a rate of 44,100 samples per second. This is good enough to reproduce sound that covers the full range of human hearing from 20 to 20,000 Hz.

Old computers and consoles were not so capable. It wasn't until around 1992 that computers could start to play samples that could theoretically match CD quality, and even then they were hampered by a lack of memory to store high quality data. Prior to that, some chips could play back 16-bit sounds at lower frequencies with very limited memory available, and even before that some could play 8-bit sounds. While those chips were explicitly designed to be able to play samples, some of the earliest chips, which according to their specification could not play samples at all, were indeed cajoled into playing back digital sounds. How was this possible?

Firstly, let's look at the SID chip. The SID was not designed to play samples, and yet in the words of the immortal Jeff Goldblum, "Life, uh, finds a way". The original SID chip has a bug in it that causes there to be a DC voltage at the output when there is no sound playing. This voltage can be increased or decreased by setting the SID's 4-bit master volume, giving a range from 0 to 15. This, in effect, becomes a simple digital-to-analog converter. By streaming numbers into the volume register fast enough, we can use it to play a sample on top of all the existing chip features. Also, the numbers can be streamed using little processing time, allowing samples to be played during a game. Many years later, ingenious new techniques were invented that have allowed the SID chip to play back multiple 8-bit or 16-bit samples simultaneously with incredible efficiency (although the SID itself has an output limit of 12 bits).

The SID had an accidental built-in 4-bit sample playback mechanism, but samples were also played on much simpler devices such as the ZX Spectrum and PC piezo speakers. These devices had an audio output that could only be on or off. If you switched between on and off fast enough, you would generate a square wave at half the frequency of the switching rate.

That's why their speakers were often called "beepers". They were designed only to play a simple beep, a single square wave generated by 1-bit data. As previously mentioned, the greater the range of values in the sample data (i.e. the greater the bit-depth), the less noisy your audio is. If you try to simply reduce a sample to 1-bit values, you will just produce a lot of scratchy noise. It therefore seems impossible to play samples at 1-bit, like trying to draw a picture using a light switch. But it is actually possible, thanks to something called delta-sigma modulation (and its cousin pulse-density modulation).

Ideally, we need to output a smooth waveform rather than a scratchy 1-bit noise, so we need to find a way to output the intermediate values between 0 and 1.

Delta-sigma modulation allows us to take advantage of the fact that a speaker cannot instantly teleport from a 0 position to a 1 position. It takes a tiny amount of time to move. It's possible to change the value from 0 to 1 and then back to 0 again before the speaker has had a chance to reach the 1 position.

By carefully timing when we switch to 0 or 1, we can output waveform positions that aren't otherwise possible.

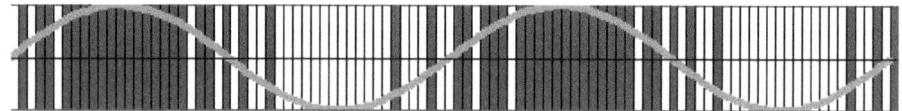

Careful toggling in a physical system can generate a regular waveform

Think about the heating system in a house. The heating element is either on or off. If you choose a specific temperature, the heater switches on, then when the temperature goes above your choice, the thermostat switches it off. When it goes below your value again, it switches it back on again, and keeps repeating the on/off cycle to keep the temperature close to your choice. Using only on and off, you can get a full range of temperatures.

Another good analogy is the *Helicopter Game* on the original iPhone, which inspired *Flappy Bird*. You have only one button for control. Hold the button and the helicopter goes up, let it go and the helicopter goes down. It's a 1-bit control, exactly like a 1-bit beeper. The helicopter is always moving forward, and you must navigate past obstacles that can be anywhere vertically on the screen by repeatedly and quickly pressing on and off. This is a type of delta-sigma modulation being performed manually by a human.

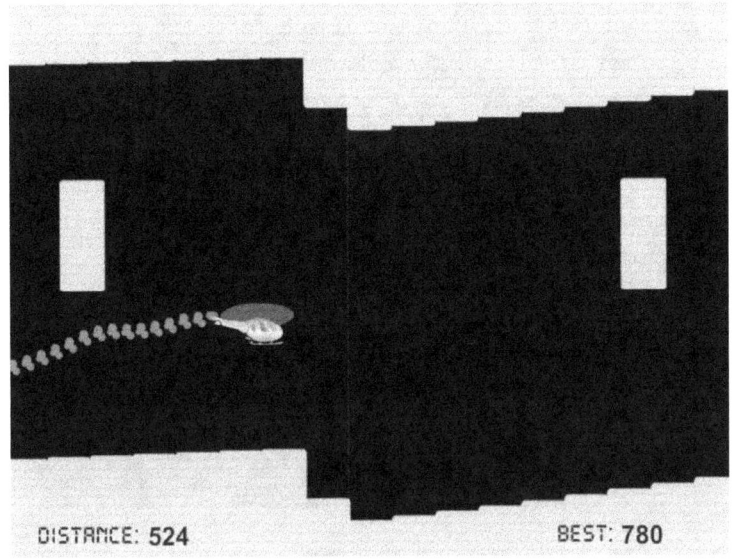

So Flappy Bird wasn't original? I'm shocked! SHOCKED!

Toggling between 0 and 1 quickly enough to play samples generates a lot of unwanted high-frequency noise, which needs to be filtered out (e.g. using a low-pass filter) to give us the actual waveform we want. In our examples, the effect of gravity on the helicopter, the rate of temperature change in the house, and the time it takes for the speaker to move into position are all examples of such a filter.

The ZX Spectrum and PC beepers are small and move incredibly fast, so the filtering is weak, and the resultant samples are scratchy. The others are incredibly slow, so trying to use those methods to play a sound would only allow you to play extremely low frequency sounds.

A ZX Spectrum replacement speaker. Beep.

I once played samples through the rumble feature of Nintendo's Wiimote. Because the rumble takes a relatively long time (compared to a speaker) to start up and slow down, it worked as a severe low-pass filter, so it could only play low frequency sounds like a car engine effect. It worked well for our racing game.

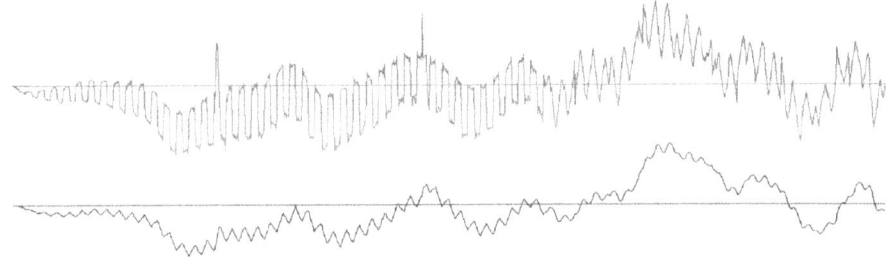

Low-pass filter affecting a waveform

Above is an example of a low-pass filter in action. At the top is the original signal. At the bottom is the same signal through a low-pass filter. The sharp spikes and corners in the original signal are the high frequencies. They have been removed or reduced by the filter, leaving only lower frequency content, which appears smoother. For the best quality audio, you'd ideally need to be able to toggle the 1-bit value at least twice as fast as your intended sample rate, and you'd need to use a filter that is tuned to filter out just the right amount of unwanted high frequency noise.

The supposed replacement of the Compact Disc, the Super Audio CD, stored audio data as 1-bit pulse-density modulation but played back the stream at a massive 2.8224 MHz, 16 times faster than the intended sample rate of 176.4 kHz.

Old processors and sound chips could never implement such a high-quality specification and playing samples with this method was very processor intensive. If you play any game on the ZX Spectrum that has speech, you'll notice that the screen always freezes while the speech is playing because the processor simply hasn't got any time to do anything else.

And even with the processor fully dedicated to playing a sample, what we're left with are low-quality sounds that are very noisy. However, those sounds were like magic and amazing to hear back in the day.

(1979) RCA CDP1863/4

A very early and very basic single-voice square wave generator with very little to recommend it.

As seen in: VP-111/VP-711 and pinball games such as Mad Race and Antar.

Single voice SSG. Mono.

Graphic chip audio bolt-ons don't come more basic than this single-voice tone generator.

Programming VIP

Its main deployment was in the $99 VP-111, known as the Cosmac VIP.

It was also included in the even-more-VIP VP-711 — a heavily upgraded VP-111, with all the optional boards and 15 more type-in listings to thrill, amaze, and probably debug!

The frequency range of the chip covers 107 Hz (A2) to 13672 Hz (G#9), but in equal 256 tone steps that weren't very musical.

The optimistic marketing offered (for the super price of $49) the **"Super Sound Board - Turns VP-111/711 into a music synthesizer! Two independent sound channels. Outputs to audio!"**.

Since it already had a tone generator, this add-on board merely added another chip of detuned frequential wonder.

Contrary to the marketing material, none of the available options were very "super".

Pinball proposition

It's a wonder this chip ever made it into the arcades, but it did in a limited number of pinball machines.

They didn't necessarily need amazing sound, just flexible sound effects, and the video part of the chip was used to generate the score display.

Computers change lives

It's easy to look down on the humble technology of 1979, but it was still mind-blowing and life-changing, as one Antipodean fan reminisces:

> *"The Cosmac VIP was our first computer and my brother and I spent many hours programming it and having fun. Living in Australia, my dad had it shipped from the US (via my uncle), and we assembled it one weekend.*
>
> *My brother learned the assembly language, and he hacked the video subroutines and created some pretty interesting effects. I am sure it was not too good for the monitor, but one effect was to get the top of the display to fold back in itself (at least that is what it looked like). Then he made a terrain type routine that looked like the ground curving onto the screen from the top and disappearing off the bottom. We added RAM to the board, we had one of our friends burn the CHIP8 language into a ROM and had the ROM dual boot into two different load states. (I think one had the CHIP8 loaded, and the other didn't). We got hold of the ASCII keyboard but never really did anything with it. It was a pretty cool keyboard and used a membrane rather than keys — very hard to use as there was no tactile feedback at all.*

I used to take the Cosmac and a monitor to school, and we would play it during recess. One of the games I wrote was a car racing game and, to work around my limited knowledge of sprite collision detection, I would draw one frame with the cars on it and test for collision, then draw another frame with the cars and scenery which did not detect for collision. As you can imagine, the scenery frame flickered quite a bit, but overall the game went well, and all my friends were impressed."

Jason Panosh (Alfredton, Australia) in a post on old-computers.com

(1979) Williams HC-55516

Pioneering quality speech chip with an incredible lifespan. Emulation doesn't do it justice.

As seen in: Gorgar (T-1), Williams System 11, N.A.R.C.

Single-voice sample playback specialised for speech synthesis. Mono.

The first to speak

The first arcade game to feature speech was not Stratovox's resource-intensive DAC-driven sample playback, as is sometimes asserted. It was Williams' *Gorgar,* a pinball machine, sporting a revolutionary HC-55516 chip.

The chip boasted seven pre-sampled words ("Gorgar", "speaks", "beat", "you", "me", "hurt", "got") that could be combined to encourage (or, more commonly, insult) the player. The sample playback/encoding was based on continuously variable slope delta modulation (CVSD). This is a technology originating from 1970, regarded as better quality than Stern's "cheap" synthesis used for their landmark *Berzerk* (although that wasn't cheap!)

Hand-crafted analog

The best description of this chip and its output is "hand-crafted". Its analog circuitry tweaks and massages the digital output to a smoothness that emulators have yet to touch.

The incredible lifespan of this chip and its successor the HC-55536 is proof of its quality.

While most of its life was spent in pinball machines, it did venture out in coin-op form in games such as Williams' *N.A.R.C.* It was used for speech and effects, usually deployed alongside a YM2151 for the music and the occasional DAC to beef up the YM2151's drums.

One notable deployment was in the Victor 9000 / Sirius S1 business computer, conceived by Chuck Peddle, who also designed the first Commodore PETs.

The S1 could use the chip to sample *and* replay sounds thanks to the chip having both encoding and decoding capabilities. This is quite amazing for a 1979 chip in a 1982 computer.

I bet the *Berzerk* designers at Stern would have used this chip if they could!

(1979) Atari C012294 (POKEY)

Four-voice PSG with a sound as quirky as its design.

As seen in: Atari 8-Bit Series (HCS) and arcade games galore!

**4 square-wave/noise channels with Atari® distortion, and a 17-bit linear feedback shift register for noise.
8-bit frequency registers.**

A crush on square waves

As you would expect from the successor to the Atari TIA, designed again by Jay Miner, the POKEY is not straightforward to understand or program.

It even develops the same quirky crush-the-waveform/distortion principle as the TIA does, but on steroids!

The "distortion" feature of the POKEY on any of these voices leads to startlingly unpredictable effects. Buzzy tones, for example, might sometimes be used to cover up note tuning issues.

What this means is that the sound of the POKEY, whether it be grindy or detuned, is distinct and unique. If you're trying to get four notes in tune, it's also a little annoying, though detuned square waves seem to be the "Atari sound" at this point.

Distortion 10	Square wave, pure tones.
Distortion 2	Triangle wave, bell tones.
Distortion 12a	Sawtooth wave, buzzy tones.
Distortion 12b	Sawtooth wave, non-buzzy tones.

What's behind the Atari sound character?

The specs of the chip reveal four semi-independent square wave generators that can vary the length of the "on" bit of the wave (the pulse width/duty cycle). This is something the AY-3-891x and SN76489 families of chips can't do, but the 6581 SID can and creates a fizzy, interesting sound. However, most musicians who programmed for multiple platforms found a way to program pulse width modulation into their driver even if it wasn't supported natively by the chip itself.

If all four voices are used, the POKEY has a similar note-tuning problem to the TIA, albeit less severe. On a TIA, there were 32 possible notes. On a single POKEY voice, there are 256, which represent evenly spaced frequencies from the top to the bottom of the range — a linear scale. But note spacing isn't linear. There are bigger gaps in frequency between notes at the top end than the bottom end, and there's no "master oscillator" to adjust where the bottom is (like the Bally Astrocade chip).

So, if you sweep a POKEY voice from 0 to 255, you will hear smooth transitions between notes at the bottom end and progressively more gaps between notes at the high end. the frequencies you need for musical notes do not necessarily fall neatly into one of the values you can access.

For sound effects, none of this is a problem. In fact, the increasingly chunky jumps as you get higher in pitch seem to add urgency and speed to certain sound effects. Also, if your jingles are simple and pitched in normal musical ranges around middle C, the out-of-tune effect is less obvious.

When two become one

For "proper" music you need more precision than 8 bits. Atari designed the chip so you could join voices 1 and 2 or voices 3 and 4 in a "16-bit" mode, combining two voices and massively increasing precision. You lose a voice in this mode, but you can accurately play any note.

The decision can be made separately for each voice pair. An article in *COMPUTE!* recommended that you might leave two of the voices in 8-bit mode for the bassline and drums and join two of them together to make a more powerful "lead" instrument, which then leaves the chip at three voices, the same as other contemporary chips. This is sensible advice if you need lead precision, though it leaves a big hole where chords could be.

For the ultimate Atari 8-bit programming guide, which really gets under the hood of programming POKEY (and everything else Atari), check out *De Re Atari* (1982) by Chris Crawford et al, available at www.atariarchives.org/dere/

POKEY at home

It took quite a while after the 1979 launch for games programmers to start to program polyphonic music on it, and even longer for them to program an original tune, musicians still being mostly absent from game companies (for instance, Atari's first dedicated musician was Earl Vickers and was hired in 1982).

The earliest POKEY polyphonic title tune I could find was in the game *Darts* by Thorn EMI (mid-1981), and it was a cover of "March Militaire No. 1" by Franz Schubert. Impressive at the time!

The floodgates opened from 1982 onwards, with tunes from Atari themselves, Synapse Software, Electronic Arts and Activision being particularly notable, and musicians such as Russell Lieblich (*Master of the Lamps*) and Roy Glover (*M.U.L.E.*) producing exceptional original scores within the limitations of the chip.

LucasArts raised the bar further with game soundtracks such as *Ballblazer* using a procedurally generated lead line. To some, these tunes sounded better on the POKEY than the SID. *M.U.L.E.* made the most of the four voices available, despite the inevitable detuned feel. Fans came to love the detune!

By the time SID musicians such as Rob Hubbard, Ben Daglish and David Whittaker migrated over to the Atari in 1986, the POKEY was usually the recipient of a ported version of a C64 master track created with a driver like that used on other computers.

As with the AY and SID chips, a critical mass of creators targeting the chip resulted in high-quality emulators and a collection of preserved chiptunes (in this case, SAP files). Later megademos and research pushed the chip further than was ever thought possible, even including code that enabled it to emulate many aspects of SID itself. The POKEY SID player is a software emulation that pushes the sound out as PCM from one of the combined channels.

POKEY in the arcades

In November 1979, the POKEY started to ship in the new Atari 8-bit machines (HCS), initially in the 400 and 800 and later in the 600XL, 800XL, 1200XL, 65XE, and the 5200 console.

It didn't take long before arcades received their first POKEY-enhanced cabinet, the legendary *Missile Command* (1980).

Having a powerful sound chip wasn't always welcomed by the programming team, who preferred sound to be a problem for the hardware engineers. Atari's Jed Margolin recounts:

> *Missile Command was the first coin-op game to use POKEY. Up until then, all the sounds in the games were done in hardware, which meant that the engineer developed the sounds.*
>
> *The Missile Command engineer didn't want to develop the sounds, so he put in a POKEY.*
>
> *Since the POKEY is programmed by software it now became the programmer's job to develop the sounds. Unfortunately, the programmers didn't want to do the sounds either. They wanted to concentrate on the game play, so they did the sounds last. By this time, management would be screaming for the game so the programmers generally adapted sound that had already been used in other games. That is why most of the games ended up sounding the same."*

Battle Zone also became a battle zone, as Jed continues:

> *Because POKEYs were bought in such enormous quantities by the Consumer Division they were really cheap, so we were told to use them instead of the more costly hardware sounds. I had just done hardware sounds for Battlezone and was told to take them out and put in a POKEY. Because of all the work I had done on the hardware sounds, I was really pissed [off] so I agreed to put in a POKEY if I could keep the hardware motor sound. Otherwise, they could do it themselves. They preferred that I do it, so I got to keep my motor."*

More coin-op deployments followed. Increasing sophistication of game requirements meant that two more versions of POKEY were created in dual-core and quad-core flavours (the original chip was labelled C012294, and the variants were C012294-02 and C012294-04).

Jed remembers this transition:

> "Star Wars originally had one POKEY and a TI speech synthesizer, which were run by the main processor. But then Greg and Norm used up all available memory (64 KBytes) and had no room for sounds. At this point in the project, it was easier to design a new board and hang it on the existing board than it would have been to redesign the main board to use memory bank switching.
>
> That is how I got to be the first to use a separate processor in an in-house developed game. Because the board had to be at least a certain size, I would have had a lot of space left over. So, I put in four POKEYs."

This was handy for having four/eight voices of 16-bit precision, and to provide stereo. Or it could just provide a boatload of sound effects. Later mid-'80s releases, such as *Marble Madness* and *Gauntlet*, kept a POKEY onboard but also added Yamaha FM/PCM chips to suit, and a TI Speech synthesizer when appropriate.

Atari used POKEY chips in some machines up to *Tetris* (dual POKEYs only!) and in *Toobin'*, *Vindicators* and *Vindicators Part II* (with those three featuring the YM2151 first heard in *Marble Madness*).

The End

> *The last game to use POKEY was the coin-op version of Tetris. By that time, Tramiel wouldn't (or couldn't) sell us any POKEYs. Things got so desperate that, to finish the run, we were paying a bounty of $50 per POKEY."*
> **Jed Margolin**

A 1988 bootleg of Atari's Tetris used four SN76489A chips instead of two POKEYs. This was quite a lot of hard conversion work for a bootleg!!

(1979) Texas Instruments SN76489

Competent and cheerful sound chip that served us well without ever blowing our minds.

As seen in: TI99/4A, ColecoVision, BBC Micro, many arcade games.

3 square-wave voices. 1 separate noise channel with two types of noise from a 15- or 16-bit Linear Feedback Shift Register. 10-bit pitch registers.

The SN76489 was the chip that, when recorded to tape in a Manchester arcade in 1983, launched Martin Galway's composing career. He once said of this chip, *"it's sort of a little cheerful chip that wanted to be cool."*

Martin recalls:

> *"It was a place on Oxford Road called Pleasure Pastimes. There were two or three arcades in that area since the times of Space Invaders. One by one they closed. Pleasure Pastimes ended up being a place known for the fact that they didn't have anything new, but... some people like to just play the old ones."*

First Texas, then the world!

A common thread in the world of video games is that the music is often more memorable and/or successful than the game. In the case of the SN76489, the sound chip was far more successful than the computer it launched in.

At the time, it was named TMS9919, and it was intended for the ill-fated "16-bit-but-not-really" TI-99/4 and TI-99/4A. This was a machine launched at $1,195 in 1979, and later sold at a loss for $49 while competing with Commodore's VIC-20. It gained some fans but lost a substantial sum for the company.

Despite the eventual grisly commercial fate of its parent computer, the SN76489 and its siblings found break-out success as an inexpensive, off-the-shelf sound generator for other systems, occasionally accompanied by their brother, the TMS9918 video display processor (for example, in the ColecoVision).

It's tough being #2

This chip was very much the Pepsi® to the AY-3-891x's Coke®, destined to always be its competitor but perpetually in second place. While the SN was a good product with a reasonable commercial success and lifespan, the SN family's public-facing "team" for home micros didn't have the AY chip's profile:

1979	TI-99/4A	TMS9919
1981-1986	BBC Micro Family (not Electron)	SN76489/A
1982	Coleco Adam/ColecoVision	SN76489/A
1982-1983	VTech CreatiVision/MK-II	SN76489/A
1983	Bandai RX-78	SN76489/A
1983	BIT Corporation Bit90 (ColecoVision cartridge compatible)	SN76489/A
1983	Casio PV-2000 (a pointlessly incompatible MSX Clone)	SN76489/A
1983-1984	Apricot P/Portable/Xi	SN76489/A

1983	IBM PCjr	SN76489/A
1983	Sega SG-1000/SC-3000(H)	SN76489/A
1983	Sony SMC-777	SN76489/A
1983-1984	Memotech MTX 500/512/RS128	SN76489/A
1984	Sega SG-1000 II	SN76489/A
1984	Sharp MZ-800/MZ-1500	SN76489/A
1985	IBM PC JX	SN76489/A
1985	SG-1000 Mark III	Sega 315-5124
1985-1991	HP9000 Series	SN76489/A
1986	Sega Master System	Sega 315-5124
1987	Tandy 1000 series	SN76489 or NCR8496 or PSSJ-3
1990	Sega Master System II	Sega 315-5246
1990	Sega Mega Drive/Genesis	Sega 315-5313
1991	Sega Game Gear	Sega 315-5377
1991	Sega Mega-CD	Sega 315-5548
1992	Sega CD	Sega 315-5548

| 1993 | Sega CD 2/Mega-CD 2 | Sega 315-5548 |
| 1993 | Sega Pico/Yamaha Copera | SN76489 |

It's a chunky table, but not much of a match for the AY's sales and presence, either at home or in the arcades.

Arguably the SN got its biggest audience by stealth, sneaking into the Sega Mega Drive/Genesis with the Master System emulation and occasionally being used in tandem with the more powerful YM2612 FM chip (for instance, in *Space Harrier*).

Hardware: serving suggestion only

The main macro difference between the AY and the SN is that the SN doesn't have a hardware envelope generator to change the shape of the sound over time.

However, the AY's hardware envelopes are sufficiently limited/inflexible that most sound/music programmers drive the sound envelope in code anyway.

This means sound moulding on both chips is usually done by the programmer's code, using the volume setting for a channel for piano-like sounds, organ-like sounds, or anything in between.

However, BBC Basic helps BASIC programmers with its surprisingly advanced "envelope" command, which is entirely CPU-controlled.

Therefore the same command can also be used on the Acorn Electron, which doesn't have an SN76489. The "envelope" command was also repurposed on the BBC Micro itself to control third party extensions such as speech chips.

While the SN76489 can be tricked into doing many things in software, one limitation that must be respected is the chip's more limited frequency range, which is two octaves fewer than the AY.

Its tone generators even play one octave lower than their calculated frequency to make the range more useful.

Same chip, different noise

Different versions or integrations of the SN sometimes changed (or even inverted) the operation of the noise waveform so that, for instance, the noise from a BBC Micro sounds different to the noise from a Sega Game Gear.

Because noise can be validly implemented so many ways, the noise generated by a chip is often its most distinctive feature.

Variation on a theme

The SN76489 can drive a small speaker from its output (useful for toys and keyboards) but the SN76489/A can't.

Other variants such as the SN76496 have an audio input pin, useful in arcade deployments with multiple chips and music sources because it allows the chip to be used as a mixer.

This chip's lifespan was helped along by clones such as the NEC version first deployed in the PCjr and copied in the Tandy 1000, which gave Rob Hubbard one last old-fashioned chip to target during his career at Electronic Arts.

It was also integrated into display chips, such as on the Sega Master System.

Functional clones of this chip are still available today.

How did the SN76489 variants differ?
(Thanks to MAME for this information)

- SN76489 uses a 15-bit shift register with taps on bits D and E, output on E, and uses the XOR function and a 15-bit ring buffer for periodic noise/arbitrary duty cycle. Its output is inverted.

- SN94624 is the same as SN76489 but lacks the /8 divider on its clock input.

- SN76489A uses a 15-bit shift register with taps on bits D and E, output on F, XOR function. It uses a 15-bit ring buffer for periodic noise/arbitrary duty cycle. Its output is not inverted.

- SN76494 is the same as SN76489A but lacks the /8 divider on its clock input.

- SN76496 is identical in operation to the SN76489A, but the audio input on pin 9 is documented.

- All the TI-made PSG chips have an audio input line which is mixed with the 4 channels of output, but it's undocumented and may not function properly on the SN76489, SN76489A and SN76494: the SN76489A input is mentioned in datasheets for the TMS5200.

- Sega Master System III/MD/Genesis PSG uses a 16-bit shift register with taps on bits C and F, output on F. It uses a 16-bit ring buffer for periodic noise/arbitrary duty cycle. It is assumed it uses XOR and inverted output.

- Sega Game Gear PSG is identical to the SMS3/MD/Genesis one except it has an extra register for mapping which channels go to which speaker.

- NCR8496 (as used on the Tandy 1000TX) is like the SN76489 but with a different noise LFSR pattern with taps on bits A and E, output on E, the use of the XNOR function and a 15-bit ring buffer for periodic noise/arbitrary duty cycle. Its output is inverted.

- PSSJ-3 (as used on the later Tandy 1000 series computers) is the same as the NCR8496 with the exception that its output is not inverted.

The award goes to...

This chip was also, for some reason, used in the Autocue 1500 Teleprompter, released in 1988.

(1980) Texas Instruments TMS3615N/TMS3617NS

Massively polyphonic square-wave sound chip designed for home organs that escaped into coin-ops.

As seen in: Lazarian, Monster Bash.

13-voice SSG. Mono.

Home Organ

In the early days of coin-op (before standardised boards) you sometimes had to cobble together what you could find. This led to some interesting use of some fascinating and quirky chips.

The TMS36xx series is one of those, originally built to power electronic home organs.

As with the OKI MSM5232RS, this chip uses the "feet" terminology for referring to octaves on a keyboard. Measurements such as 2 feet, 4 feet, 8 feet and 16 feet originally referred to the length of the pipes in a pipe organ. Halving the length of the pipe increases the pitch by an octave.

Think of a didgeridoo vs a flute: a big pipe vs a small pipe. The big pipe has a much lower sound and isn't capable of the same high frequencies the flute is. But there is probably a little overlap. Even with a pipe of the same radius, a longer pipe has a lower sound.

Each chip supports only 13 notes, and all 13 can be output at once if you do the digital equivalent of mashing the keys. Not only that, but output is replicated an octave lower, which gives a characteristic organ-like sound, so you can choose 16 feet and 8 feet together, 8 feet and 4 feet together, or 4 feet and 2 feet together.

13 notes aren't really enough, but they thought of that. Each chip also has an output that passes on a half-speed clock signal to another identical chip.

This halves the "footage" of that chip, making it lower-sounding. Daisy chaining the chips from high to low results in more coverage, though it's only the top chip that gets all 13 notes (because it needs a top "C" of its own).

Daisy daisy

Three chips offering 37 simultaneous notes sounds good, but if you played them all together, the sound would be unbearable ear torture. In practical terms, the music played by the chip must be musical (maybe a bassline, a lead, and some chords). Usually between 5 and 7 notes would be played at once, depending on how musically clever the game was trying to be.

The coin-op *Laser Battle* uses these chips for classical music. That game also has the *Space Invaders* SN76477 chip and a trio of Signetics chips, so it's well stuffed with audio, but the organ sound is quite distinctive.

Other games using this chip are *Phoenix, Pleiades,* and Sega's *Monster Bash.*

The latter connected its TMS36XX to an NEC N7751 CPU that was also used to directly generate samples through an 8-bit DAC.

(1980) National Semiconductor MM5837

A simple, well-engineered noise generator. Sometimes all you need or want.

As seen in: Bally/Sente's Goalie Ghost, Snake Pit, and others.

Single-voice noise generator. Mono.

When you've installed 6 musical chips in your coin-op and you realise that bullets and chickens both make a kind of white noise that the chips can't do, to where do you turn? If you're Bally/Sente you turn to a dedicated noise generator such as the MM5837.

The datasheet boasts of "Uniform Noise quality, Uniform Noise amplitude, eliminates noise preamps, self-contained oscillator".

A closer look shows it's a shift register as usual, 17-bits long with taps on bits 17 and 14, emulating white and pink noise. There are no frills — you must process the volume of the noise yourself.

In fact, if you didn't feel like buying the chip, you could program your own in software.

BYTE magazine helpfully replicated the noise-generation algorithm of this chip back in 1980 (see over), so boot up that assembler and get noisy!

Having said that, while the noise generation algorithm might be identical, the overall quality and build of the technology will always be important to the result, and quality is what the MM5837 offers.

Good engineering makes the difference between a noise and a racket!

Programming Quickies

A White-Noise Generator for the Apple II

John O'Flaherty, 3432 A Evergreen Ln, St Louis MO 63125

Listing 1 is a simple machine-language routine to turn an Apple II into a white-noise generator. The program is a software machine that simulates the National Semiconductor MM5837 Digital Noise Generator (see figure 1).

It uses 2 bytes of memory, hexadecimal locations 300 and 301 (see listing 1) as sixteen of the shift-register stages, and the processor-status-register carry flag as the seventeenth.

The rotate-left (ROL) instruction at hexadecimal location 303 shifts the bits of the low-order memory location (hexadecimal 300) left, moving bit 8 into the carry flag. The next ROL instruction, at location 306, shifts each bit of location 301 left, shifts the carry flag into bit 0 of location 301, and shifts bit 8 into the carry flag. One seventeen-bit shift cycle is now complete.

At this point, if the carry flag, which is now the output bit of the seventeen-stage register, is equal to 0, the program jumps to location 30E; but if it is set to 1, the program toggles the speaker by the instruction at hexadecimal location 30B.

Now the accumulator is rotated right three times, bringing the carry flag (bit 17) into bit 6 of the accumulator, which is exclusive-ORed (at location 311) with bit 6 of location 301 (bit 14). Then the accumulator is shifted left three times to put the bit of interest back into the carry flag. Then control branches back to address 303 with the correct bit ready to be shifted into the front of the low-order memory byte by the ROL instruction.

The routine is entered at hexadecimal address 302. Reset must be pressed to stop the program.

It is also possible to insert counting loops and a conditional subroutine return to create a time-limited burst of white noise: the program in listing 2 will produce a short "chiff" sound.

With seventeen stages of shift register in a pseudo-random circuit, there are nearly 2^{17} or 131,071 unique states. The cycle time of the loop averages 27 microseconds, so the total cycle time before repetition will be 3.54 seconds (for the program of listing 1). ∎

Listing 1: *6502 assembly language program for a continuous white-noise generator.*

```
300   XX              (low-order)
301   XX              (high-order)
302   38              SEC
303   2E  00  03      ROL   $300
306   2E  01  03      ROL   $301
309   90  03          BCC   $30E
30B   AD  30  C0      LDA   $C030
30E   6A              ROR   ACC
30F   6A              ROR   ACC
310   6A              ROR   ACC
311   4D  01  03      EOR   $301
314   0A              ASL   ACC
315   0A              ASL   ACC
316   0A              ASL   ACC
317   4C  03  03      JMP   $303
```

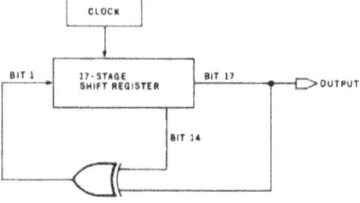

Figure 1: *Logic diagram of the National Semiconductor MM5837 digital noise generator circuit.*

Listing 2: *Subroutine to generate bursts of white noise.*

```
300   XX              (low-order)
301   XX              (high-order)
302   A9  00          LDA   #$00
304   A8              TAY
305   38              SEC
306   2E  00  03      ROL   $300
309   2E  01  03      ROL   $301
30C   90  03          BCC   $311
30E   AD  30  C0      LDA   $C030
311   6A              ROR   ACC
312   6A              ROR   ACC
313   6A              ROR   ACC
314   4D  01  03      EOR   $301
317   0A              ASL   ACC
318   0A              ASL   ACC
319   0A              ASL   ACC
31A   88              DEY
31B   98              TYA
31C   D0  01          BNE   $31F
31E   60              RTS
31F   4C  06  03      JMP   $306
```

Byte Magazine making a big noise... and don't forget to fill in the inquiry card!

Arcade Speech Synthesis - when it was interesting!

A review of early speech synthesis in the arcades

In the early days of the arcade, machines in the corner speaking to you seemed like pure sci-fi, so it was a guaranteed way to attract quarters/yen/10p coins.

> *The first time I ever heard speech was walking past a Berzerk [cabinet] and hearing 'Coin detected in pocket.' I still think that was a great bit as it had nothing to do with the game, but it sure did work as an attract."*
> **Internet forum poster**

Speech synthesis in the arcades went in a bit of a circle though. First, samples were used because the technology was too young. Then came speech synthesis, based on phonemes, formants, allophones, and physical modelling. Finally, samples returned because it was cheaper to just record and playback speech once memory prices dropped. Also, robotic-sounding synthesized speech exemplified by the Speak and Spell toy and eventually Stephen Hawking's voice generator had become less desirable.

The first arcade game to feature speech was a pinball machine: Williams' *Gorgar*, using the HC-55516. The first video game to use speech synthesis was *Stratovox*, known in Japan as *Speak & Rescue* (May 1980). Developed by Sun Electronics in Japan and making it into North America via Taito, this shooting game features several phrases such as *"help me"*, *"very good!"*, *"lucky!"* and *"we'll be back"* (or the equivalent Japanese phrases).

It used an 8-bit DAC driven by a Z80 CPU.

King & Balloon (Coin-op, Namco, June 1980)

Released within a month of *Stratovox*, *King and Balloon* also used a second Z80 processor to control a DAC for speech.

Berzerk (Coin-op, Stern Electronics, November 1980)

While some earlier games used samples played back through a DAC, *Berzerk* is one of the first video games to use genuine speech synthesis, featuring talking robots which are a natural fit for the technology!

The game used custom LPC speech synthesis encoding developed by part-time speech enthusiast and Professor of atomic physics Dr Forrest Mozer.

It was played back through a TSI S14001A, developed by TeleSensory, Inc. in 1975 as a speech chip for a portable talking calculator for the blind.

In 1980, computer voice compression was extremely expensive, and it's estimated to have cost the manufacturer USD $1,000 to digitally compress each word. Minicomputers don't pay for themselves!

> " *Forrest Mozer would encode the speech in his basement laboratory using his novel form of speech encoding. The encoding process apparently involved several minicomputers running FFTs and a spectrum analyzer.*"
>
> *Jonathan Gevaryahu, TSI S14001A Speech Synthesizer LSI Integrated Circuit Guide*

The English-language version of the game has thirty words in its robotic vocabulary.

Through Dr Mozer, there is a direct line in compression technology between this chip and the later home-computer speech in games such as *Ghostbusters* ("*Mwah ha ha ha haaaaaa!*").

Wizard of Wor (Coin-op, Midway/Dave Nutting Associates, December 1980)

Given how long arcade development takes, this machine was almost certainly in development at the same time as Berzerk since it was released only a month and a half later.

> "*The SC-01 Speech Synthesizer is a completely self-contained solid-state device. This single chip phonetically synthesizes continuous speech, of unlimited vocabulary, from low data rate inputs.*
>
> *Speech is synthesized by combining phonemes (the building blocks of speech) in the appropriate sequence. The SC-01 Speech Synthesizer contains 64 different phonemes which are accessed by a 6-bit code. It is the proper sequential combination of these phoneme codes that creates continuous speech.*"
> **Votrax® Datasheet**

If you were using this chip and didn't want to do the heavy lifting, Votrax® had you covered.

> *Automatic Operations: Votrax® can supply a micro-computer system for automatic conversion of English text into phoneme sequences. This system is particularly useful for in-house vocabulary development and product security. Contact Votrax® for further information.*"

Apart from *Wizard of Wor*, it was used in *Alien Voicebox I* and *II*, *Gorf*, the Heathkit HERO 01 robot (ET-18), Heathkit HERO Jr. (RT-1), *Q*bert*, Votrax *Type 'N Talk* and more.

But what was the system saying? *Wizard of Wor* has an astonishing 83 phrases, from *"Hey! Insert Coin!"* to *"Come back for more with the Wizard of Wor. Ha ha ha ha!"*. It even addresses you as *"Worlord"* instead of *"Worrior"*[sic] after you've reached the final dungeon.

Berzerk gets all the credit, but *Wizard of Wor* is an amazing feat of imaginative vocab. *Gorf* has only 31 phrases, but that's compensated for by an increase in sass level (*"Some galactic defender you are!"*).

Vanguard (Coin-Op, SNK, 1981)

This is possibly the naughtiest early game to steal copyrighted music (for instance, the *Star Trek: The Motion Picture* soundtrack). This vertical/horizontal shooter was SNK's first colour game also has some flat robotic speech.

The speech in the game announces the names of each level, such as *"Rainbow Zone"*. It also says things like *"Be careful!"*, which should probably have been better directed at their legal department, thanks to the music rip-offs.

Astro Blaster/Space Fury (Sega/Gremlin, March/June 1981)

Well, that escalated quickly. Speech synthesis was a growth area in 1981, and General Instrument was determined to compete with Texas Instruments. These two games feature their SP0250 Speech Board, known as "The Orator". Its aim was to provide voice synthesis with inflections and expression, something noticeably absent from its rival-to-the-death, the Votrax SC-01.

General Instrument is generous with information in its datasheet while never missing the chance to advertise its PIC1650A CPU.

> "The SP0250 speech synthesizer is an N-channel MOS LSI device capable of generating high-quality speech with the natural inflection and emphasis of the original speaker. Operation requires one or more ROMs to store speech data and a microcomputer/processor such as General Instrument's PIC1650A…
>
> The analog speech signal from a tape recording is applied to a 1-5 kHz bandpass filter and sampled at a 10 kHz rate. Each sample is converted to a 12-bit digital value. A pitch period estimation and voicing decision is made to obtain the pitch, amplitude, and repeat coefficients. An LPG analysis is then performed on the digital data which produces the digital filter coefficients to best match the spectral characteristics of the samples in a particular speech frame. During this analysis, the number of filter stages and the coefficient precision (low, high, full) information is entered. The coefficients generated by the analysis are then translated into a form which is compatible with the SP0250. If desired, the data may be further compressed by delta coding the coefficients."
>
> **General Instrument SP0250 datasheet**

Phrases that Astro Blaster threw at the player include "*Fighter pilots needed in sector wars, play Astro Blaster!*", "*Alert, alert, invaders in sector 1, player 1 to battle stations!*", "*Fuel status marginal!*" and other status messages.

Space Fury is sassier, even surpassing the notorious sassiness of *Gorf*:

> "Is there no warrior mightier than I?"

> "Does anyone dare challenge my imperial fleet?"

> "So, a creature for my amusement. Prepare for battle!"

> "So, you defeated my scouts. Well, my cruisers will destroy you."

> "You are starting to annoy me, creature. My destroyers will annihilate you."

> "You survived! Warships! Dispose of this annoyance at once."

> "Well done. Prepare to battle my entire fleet!"

Konami (1983 - 1988)

The Sanyo VLM5030 was heavily featured in Konami games such as *Track and Field* (1983), *Hyper Sports* (1984), *Gradius* (1985) and *Konami '88* (1988).

Technical documentation is lacking, though MAME's (imperfect) emulation is based on the TMS5220 chip, implying that this is a me-too attempt based on LPC.

Discs of Tron (Coin-op, Bally Midway, 1983)

The Texas Instruments family (in this case, the TM5200 VSP) is deployed on Bally's AS-2518-61 "Squawk & Talk" board installed in *Discs of Tron*. The speech data was packed into a TMS6100 VSM chip (Voice Synthesis Memory).

Atari Classics - TMS5220C (1985)

Texas Instruments had been leading the field in speech synthesis for some time, with a system good enough to be used in various arcade games years after it was released in 1980. It was used in big titles such as *Star Wars*, *Return of the Jedi*, and *Road Runner*. (You've got to wonder what the speech analyser made of "*Meep Meep*".)

One common feature between "The Orator" in the Sega machines and this chip was the attempt to mathematically model the biological equipment and processes humans use to make speech sound realistic since the flat, robotic sound got old quickly.

> *Speech data that has been compressed using pitch-excited linear predictive coding (LPC) is supplied to the VSP either by the CPU or by direct serial access of a Voice Synthesis Memory (VSM). The VSP decodes this data to construct a time-varying digital filter model of the vocal tract. This model is excited with a digital representation of either glottal air impulses (voiced sounds) or the rush of air (unvoiced sounds). The output of this model is passed through an 8-bit digital-to-analog converter to produce a synthetic speech waveform."*
> **Texas Instruments TMS5220C datasheet**

LPC is still used as the core of most digital speech compression algorithms, including for digital cell phones.

Coin-op titles using the chip during the mid-80s included *Indiana Jones and the Temple of Doom*, *720°*, *Gauntlet*, *Gauntlet II*, *A.P.B.*, *Paperboy*, *RoadBlasters*, *Vindicators Part II*, and *Escape from the Planet of the Robot Monsters*.

> "All the '80s Atari games that used speech used the TI LPC speech chip, except Cyberball (68000 processor, sampled audio) and Xybots (used the Yamaha music chip to do speech as well)."
> **Earl Vickers, legend, ex-Atari**

After 1985, hardware performance had increased so much that nearly all games used simple sample reproduction from then on.

(1980) Commodore MOS 6560/6561 (VIC)

Much-loved, idiosyncratic, and clever four voice SSG with a raw, distorted sound.

As seen in: VIC-20/VIC-1001/VC-20

3 square wave voices with varying frequency range (bass, alto, soprano). Approximately 128 notes available in total.
1 noise waveform. Noise channel is 16-bit LFSR with taps on 3, 12, 14 and 15.

This was a solid start for chip designer Bob Yannes, giving a taste of the incredible adventures to come. There's no SID without VIC!

Masses, not classes

By 1980, sound chips in the wild were deployed in games consoles (Intellivision, VCS), coin-ops or pricey home computers such as the Atari 8-bit range or the TI99/4, both of which had disappointing sales until they were reduced in price.

One of the reasons why these machines weren't pushed harder musically and sonically was that the intersection of musicians and programmers was very small indeed.

This was partly a numbers game — home computers just hadn't reached that many homes. In addition, companies like Atari and Texas Instruments were keeping a tight control on software released commercially for the machines, in direct contrast to the flourishing-yet-mostly-beepy Apple-II ecosystem.

The VIC-20 was tossed into this cozy picture like a grenade, albeit one filled with the raw essence of post-Star Trek William Shatner in its advertising.

The resulting price war with TI and Atari was calamitous for both of those companies, but at least the resulting price drops on their machines meant that home computers ended up in many more hands. This was good for consumers, bad for company profits. Texas Instruments maintained their iron grip on commercial releases to the end, even while shipping out millions of loss-making machines. However, Atari loosened their grip in 1980 with the "Atari Program eXchange", resulting in games like *Eastern Front* and *Caverns of Mars*. More importantly, they also released the technical documents of the 400/800, so third-party software started appearing in volume from 1981.

The VIC-20 was a huge success. Not only was it cheap and loud, but it was internationally accessible.

It was even sold at retail outlets such as K-Mart (the popular TRS-80 also achieved success at retail, being sold at Radio Shack). Commodore encouraged third-party development, and the technical documentation was excellent.

This "masses not classes" approach increased the number of creative minds being exposed to a home computer for the first time. Even if the graphics and sound were basic, and the BASIC was also basic, it got the juices flowing — the VIC-20 became the first home computer to sell over one million units worldwide.

Code and quirks

Sound was very much a second-class citizen in VIC-world, borrowing a small corner of the video chip's silicon. There were no BASIC commands or convenience functions to access the sound features, for instance.

Commodore BASIC (famous for driving people to code in machine language) made learning how to PEEK and POKE cryptic addresses a vital part of the creative process.

The most obvious quirk of the VIC is that the voices each have different octave ranges, equivalent to bass, alto, and soprano, with a limited number of possible notes within each range.

This was a reasonable solution to the note-precision problem highlighted with the Atari machines, especially given the lack of ambition of most musical compositions on the VIC.

The waveforms were also very distinctively shaped. The square wave doesn't look very rectangular at all, and the noise generator is very, very quirky (see the Super Tech Talk section on noise generators to find out why).

Loud chip, quiet games

The VIC was not really pushed by the games released for it. They didn't have the memory or (in 1980 - 1981) any commercial reason to feature polyphonic title tunes. Continuous in-game tunes also required advanced interrupt-driven code. Brash jingles and sound effects were the order of the day, which the quirky noise generator handled well.

The best use of the VIC's sound was Jeff Minter's game *Gridrunner*, sounding better than the version on the Commodore 64 because of the rough waveforms and tuned noise.

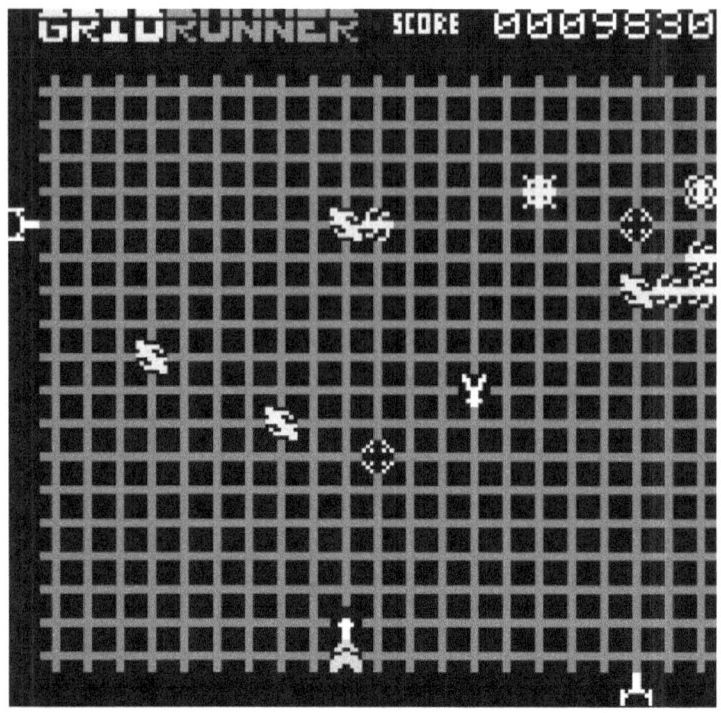

Jeff Minter's Gridrunner was a VIC legend

Most games were far less ambitious. For instance, an action game called *Attack UFO* was a *Space Invaders* clone with a series of dog whine noises and pffts — a wasted opportunity, indeed.

A very modern development

As you'd expect, the VIC has been pushed to (and beyond) its limits with recent megademos such as *Robotic Liberation*, thanks to modern chipmusic techniques and a variety of external development tools. There's even a SID player, though that ignores the VIC voices in favour of using the chip as a PCM output device.

(1980) Namco WSG/52xx/54xx/15xx

Iconic wavetable sound generator that defined arcade sound for years.

As seen in: Pac-Man, Galaga, Pole Position.

3 wavetable voices. 16 waveforms, 32 x 4-bit entries. Waveforms stored in external 256-byte PROMs.

Human ears had never heard this before

The popularity of *Pac-Man* means that the original Namco WSG, more than any other, entered the cultural zeitgeist as "the sound of video games" and never really left. When *Pac-Man* was released in 1980, home computers and consoles were yet to touch the lives of most people, so the *Pac-Man* "wakka wakka" chomping and "woo woo woo" siren noises became the definitive arcade soundtrack (along with the *Space Invaders* pulse and shooting noises). Both games and their sounds have been endlessly parodied.

Even though the sounds are easy to parody, they're difficult to emulate. Namco's innovative wavetable design made them hard to emulate with the SSG chips on home micros, proving a challenge for accurate arcade conversions.

This was the first sound chip to implement fully featured wavetable synthesis. Instead of using mathematically based, regular waveforms like square waves, it implements 8 small snippets of waveform per memory chip that are playable at different speeds to achieve music or effects.

This waveform generation method has advantages and disadvantages. The waveforms can be anything, allowing unique sounds. However, there isn't a lot of space for the waveforms, so they must be very short.

Waves and steps

These particular waveforms are 4-bit samples with 32 "steps". Each step in the sequence is an integer value from 0-15, defining how loud the sound should be at that point. Played in sequence (and usually looped), these humble numbers are the basis of sonic magic.

The chip doesn't have an oscillator to produce waveforms such as sine, triangle or sawtooth. This means any of those wave shapes needed for the game must be implemented as jaggy (but serviceable) user samples.

The 8 waves are stored on external 256-byte PROMs. Each chip could be refilled with different samples from game to game. *Pac-Man* had two of these chips, using 12 of the 16 possible waveforms, though no one seems to know if samples 9-12 were ever used. There is data there, but it doesn't seem to be used during gameplay.

Getting the right effect

A sound effect in *Pac-Man* is a sampled waveform and some modulation instructions, such as "play the triangle waveform in a loop and quickly slide the pitch up and down to create a siren effect".

Namco gave voice one improved frequency resolution (20-bit). The other two voices have respectable 16-bit frequency registers. This results in a different frequency table for voice one, meaning it's incompatible with effects designed with the frequency tables for voices two and three.

That music

For the music, the main *Pac-Man* jingle at the start of a level is composed for two-voice polyphony. Voice 1 is lead, voice 2 is bass.

The iconic Pac-man level start jingle.
Transcribed by David Youd, copyright Namco.

Custom Chip Shop

Namco's next coin-op game, *Pole Position*, had additional sonic requirements, so Namco had to deploy two more custom chips, the 52xx and the 54xx.

Both are actually Fujitsu MB8843 MCUs (microcontrollers) with mask ROMs: that is, they are general purpose CPUs that have been pre-programmed for specific tasks.

The 52xx provides 4-bit sample playback of samples that are too large for the WSG's 32-byte limit, such as crowd cheering and phrases such as *"Prepare to Qualify"* and *"Good Driving, You've Qualified to Race!"*

The 54xx is a sophisticated 3-voice noise generator with a filter to change the character of the noise, able to generate anything from a rumble to a screech.

In *Pole Position*, it's only used for tyre screech noises and the explosion when you crash. These noises are also beyond the scope of the main WSG's synthesis methods.

The Namco 15xx that was put into the *Super Pac-Man* board (used also for *Dig Dug 2*) increases the number of music voices from 3 to 8, but the samples are the same size, and stored in the same way.

Despite being Namco-specific (and mostly famous for being used in classics such as *Galaga*, *Pole Position* and *Dig Dug*) this chip also saw limited use in non-Namco games, such as Sega's *Pengo* and *Ali Baba and the 40 Thieves* (both 1982).

Pengo-man

> *One of Pac-Man's sprites, the sixth frame of his death animation, can be found within the game's graphic banks. This is the only remnant of Pengo being built off of Pac-Man's hardware."*
>
> **The Cutting Room Floor website**

Midway games were using *Pac-Man* hardware as late as 1985 (*Jump Shot*).

Pic-Man, Crock-man and Joy-man, man

Namco's lawyers were kept very busy as many bootleg games were based on Namco's *Pac-Man* hardware, with some implementing software emulation of the WSG, which avoided having to obtain the chip from dubious sources.

The most notorious of these was Midway's own (initially) unauthorised *Ms. Pac-Man*, but there were numerous clones such as *Crock-man* or *Joyman*.

Pac-Man bootleg cabinets collectively sold as many units as the original machine (approximately 300,000).

Ghost meets Pac. There can be only one winner.

WSG Family Waveforms

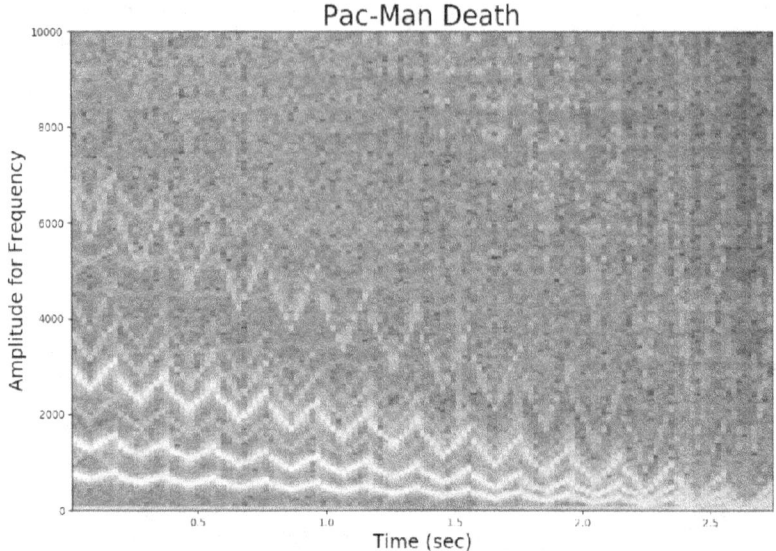

This is what death looks like.
This graph shows the sound visually. The bright lines show the pitch. In this case, a gradually decreasing series of upward slides terminating in a splat.

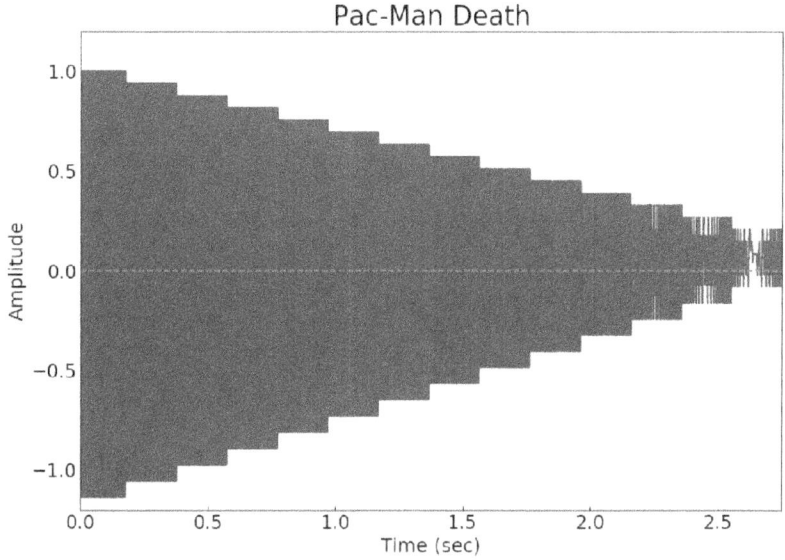

Although the waveform seems to get quieter,
to the listener, it gets lower in pitch instead.

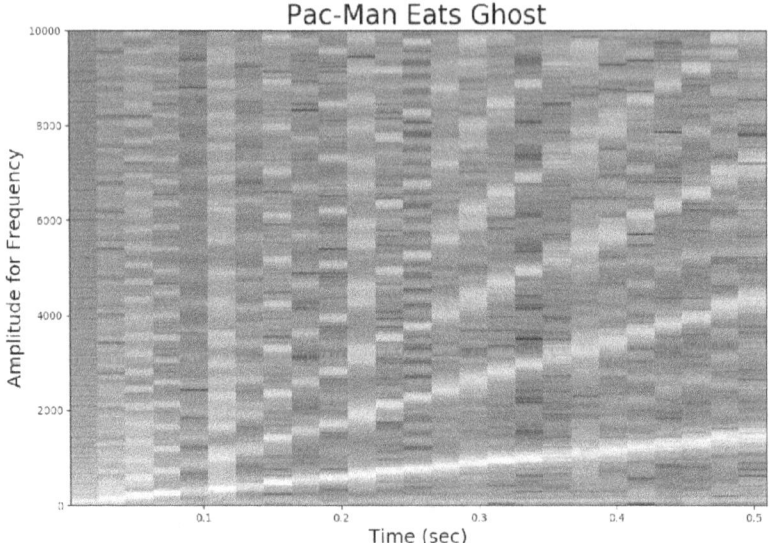

Spectrogram shows a noisy upward slide (bright line), with additional harmonics (dimmed lines).

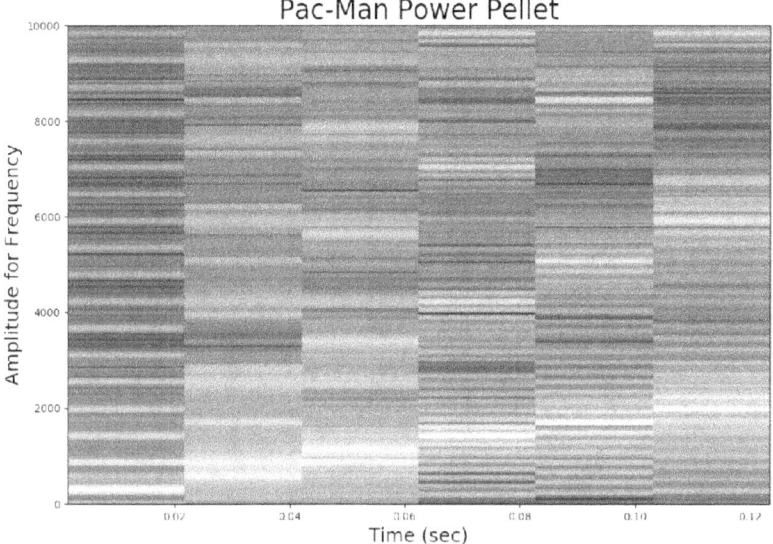

Spectrogram: follow the bright line for the pitch. You can see how this sound moves in steps upwards approximately every 0.02 seconds.

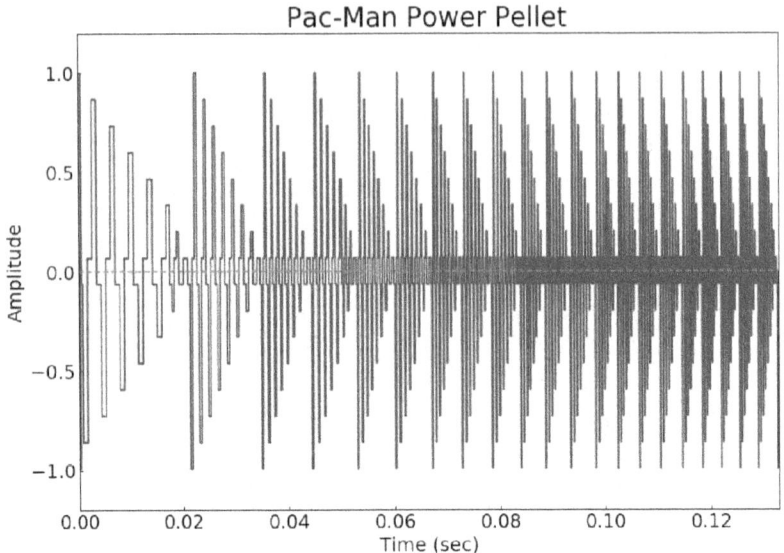

From this graph, you can see how the sound is square-wave based.

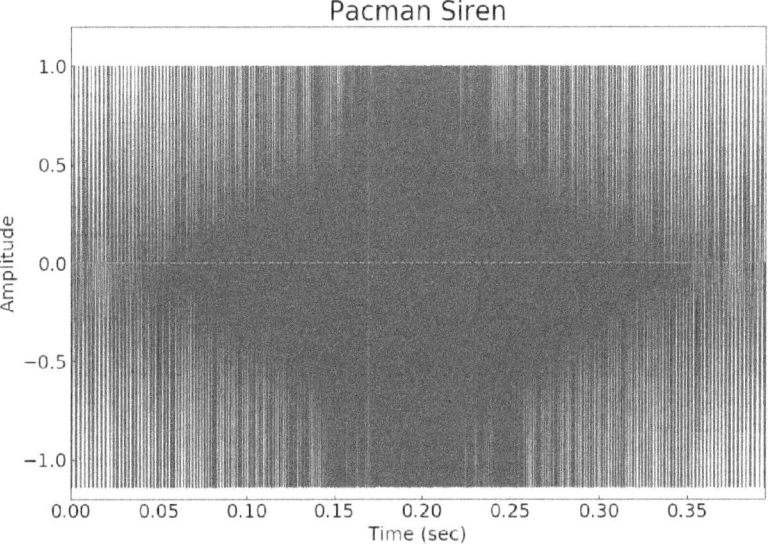

Background triangle wave going up in pitch, then down, before repeating

(1980) RCA CDP1869

Underwhelming and obscure two-voice SSG.

As seen in: COMX-35, Pecom 32, Cidelsa arcade games

2-voice SSG. Mono.

Two whole voices?

RCA, the heroes of the VP-111 and its 256 notes return, this time with two voices: one square wave and one noise.

While this chip is a definite upgrade from the previous one, its frequency register is only 7-bit, rather than the previous 8-bits. This might not seem like an improvement, but...

A trick of the range

On the previous chip, the range covered by the frequency register is spread over the entire audio spectrum, from "low bass" to "annoy dogs". However, most music stays within a much smaller range, as you can tell by listening to the charts these days, where two notes often suffice.

This chip allows you to limit the top frequency it can address, making better use of the available range. It promises "576 selectable tones over 8 octaves".

This is acceptable pitch accuracy.

Regional computer love

The main use for this chip was in the COMX-35 home micro, 1985's Serbian Pecom 32 and its 64K upgrade Pecom 64, and the Finnish Telmac TMC-600 home computer (only 600 units produced!)

The COMX-35 was launched in several countries but seems to have been most popular in the Netherlands, the most notable software being a port of *Boulder Dash*, which has acceptably diggy noises, but no title music. The launch in the United Kingdom was hampered by quality control issues and competition from more established home micros such as the Amstrad CPC.

The Pecom computers give a window into how countries behind the Iron Curtain fared with IT literacy:

> *"Those computers were built in demand from schools.*
>
> *We got this computer in our school to learn BASIC (at that time ex-Yugoslav government had 5 different computer projects).*
>
> *The programming language was BASIC with no real graphic commands. The year after, in our school, all those Pecoms were replaced with IBM PC clones. The size of the computer was a little bit smaller than an A4 page. It had tape as primary media…*
>
> *I don't remember the OS, it was home-made, based on CP/M (as were many systems at that time in Yugoslavia). Price? Not so cheap, it was 2 average monthly salaries."*
> **Darko Sola, Pecom fan**

The COMX and the Pecom are both emulated with the *Emma 02* package, a tribute to the eternal love of people and their first computer!

Clone City, Arcadeville

The chip found use in the arcade from the pioneering yet paradoxically unoriginal Spanish manufacturer Cidelsa.

They put two chips in each machine for extra polyphony in games such as *Destroyer* (a *Phoenix* clone), *Altair* (a *Galaxian* clone) and *Draco* (a *Berzerk* clone).

Speech Synthesis - Home systems

A look at speech synthesis for home microcomputers.

Limited vs Unlimited

Speech synthesis has two basic categories: unlimited vocabulary and limited vocabulary. To be able to offer unlimited vocabulary, a speech synthesis chip must have some of the basic building blocks of language on board and a way of triggering them in a particular order to sound as natural as possible (if speech other than a shout of "Ghostbusters!" ever sounded natural).

At this stage of speech synthesis technology, unlimited vocabulary necessarily involved flat-sounding, robotic voices. Advanced off-line processing was required to convert text into realistic, expressive speech, involving complex analysis and reconstruction of particular words based on inflection, pitch, and stress, while modelling the various physical/biological processes involved.

It also required sophisticated compression and encoding to minimise the size of the speech.

Speech synthesizers didn't have the space to store how every dictionary word was pronounced, which meant that it had to be told. An unlimited vocabulary therefore required the user to type words foanetikallee (i.e. as it sounds, not how it's written).

The Voice Software Speech Synthesizer for Apple II (1980)

In 1978, Bob Bishop developed an Apple II program that repurposed the cassette port hardware into a 1-bit sampler that could record audio for playback through the Apple II's 1-bit sound.

In 1980, Silas Warner built off Bishop's approach and developed *The Voice*, which used sampled phonemes to produce speech. Warner's bigger claim to fame came the following year with his *Castle Wolfenstein*, the first 8-bit game to include digitized speech ("*Halt!*", "*Achtung!*", "*Kommen Sie!*").

Texas Instruments TI-99/4A Speech Synthesiser (1981)

TI's TMS5200 is a classic in speech synthesizer terms, with its clever linear predictive coding, making the decoding process efficient and simple. This results in an expressive and clear (but ROM-limited) set of words. Deployed on the TI-99/4 and 4A, it was supplied free with several cartridges such as *Parsec* and *Alpiner*.

TI expert Dr Thierry Nouspikel has this to say on his TI-99/4A Tech Pages:

> *The TI Speech Synthesizer module has two custom ROM chips. They contain a 'binary-tree' list of words in plain ASCII, with the addresses where to find speech data for each word. If you want to view the content of these ROMs, you can use my Module Explorer program, and set the memory type as 's'.*
>
> *Texas Instruments originally intended to release more ROMs to extend the vocabulary of the module. That's what the little door on the top of the module is for.*
>
> *However, they realized that speech produced merely by concatenating words is of poor quality — After. All. Nobody. Speaks. Like. That!*

> To some extent, it is possible to improve it by reading speech data, modifying a frame here and there (e.g. stripping the last frame of a word to link it to the next) and feeding the result to the synthesizer. But even like this, the result is poor.
>
> TI thus created a much more versatile program that was integrated into the Terminal Emulator module (don't ask me why). The module GROM contains speech data for a list of allophones, (i.e. all possible sounds in English).
>
> The module ROM contains two subprograms: the first one breaks plain English sentences into a list of allophones, and the second creates the speech data from those allophones, adds accentuation and voice inflexion, and passes it to the speech synthesizer via the Speak-External command. This provides a much more satisfactory discourse."

This was presumably an early attempt to add accessibility features to its text-based Terminal Emulator. Very enlightened!

Intellivoice Voice Synthesis Module (1982).

Given that General Instruments created the SP0256 "The Orator" and most of the other chips at the heart of the Intellivision, the Intellivoice was a no-brainer at the time. Its predecessor had powered two Sega coin-ops, *Astro Blaster* and *Space Fury*, which were highly regarded. However, the words in the ROMs were specific to the game and did not have to be more generalised.

The ROM for Intellivoice contained phrases you'd find in Intellivision games (e.g. "*press*", "*enter*", "*and*", "*or*" and "*Mattel Electronics Presents*"). The latter phrase was presumably inserted by the marketing department.

There were games that required this module to function, such as *Space Spartans* (which stored its words on the main module), *Bomb Squad*, *B-17 Bomber*, and *Tron: Solar Sailer*. A fifth game, *Intellivision World Series Baseball*, was merely "voice-enhanced" if the module was present. The custom sample ROMs increased the cost of the cartridges dramatically, and disappointing sales for the module guaranteed its early demise.

Acorn Speech Synthesiser Upgrade for the BBC Micro (1982)

Acorn's technological foresight is admirable — the BBC Micro (even the model A!) shipped with two sockets for speech synthesizers, the upgrade having to be performed by a dealer if you didn't want to void your warranty.

The upgrade consisted of the TMS5220 chip and a TMS6100 Phrase ROM ("PHROM") that contained 165 ready-made words and word-parts. While the vocabulary was limited, the diction was world-class, thanks to the golden vocal talents of respected BBC newsreader Kenneth Kendal.

Another feature was added when you upgraded:

> *"When this extension has been fitted, you will notice that two sockets have appeared to the left of the keyboard.*
>
> *These sockets are provided to let you insert special plug-in cartridges which will be available in the future. Not only will you be able to expand your Speech System in this way, but you will also be able to use other cartridge-based software with a wide variety of applications. Note that the Speech System must be fitted to your BBC Microcomputer before you can use any plug-in cartridges."*

The most sensible upgrade at that point was to purchase the Computer Concepts Text to Speech ROM, which would enable that all-important text-to-speech profanity.

SAM - Software Automatic Mouth (1982)

This synthesizer with unlimited vocabulary for the Apple II, Atari 8-bit and Commodore 64 was created by Don't Ask Software (now SoftVoice, Inc.) The Apple version needs an expansion card (except for the hacked version of the software that sends a signal through the speaker), and the Atari version uses POKEY. The Commodore 64 version boasted the best speech quality thanks to the 4-bit DAC on the SID, and it's most familiar to Commodore 64 users from Interceptor Software's *Tales of the Arabian Nights*.

It also appeared in Gremlin Graphics' *Suicide Express*. There are tell-tale signs in *Suicide Express* that the speech engine was ripped out of the Interceptor Software game and repurposed.

The system was built around a Text-To-Phoneme converter called "reciter" and a Phoneme-To-Speech routine for the final output.

It's robotically good!

After Mark Barton created SAM, he went on to create *Macintalk*, speech synthesis for the Macintosh, as well as *Narrator*, speech synthesis for the Amiga.

The Echo 2 Speech Synthesiser card for the Apple II and Apple 2e (1982)

This is a plug-in expansion card based on the popular TMS5220 LPC Speech Chip, and capable of unlimited vocabulary (male voice) and limited vocabulary (female voice).

Mockingboard with Votrax SC-01 (1983)

This Apple II music board add-on came with an option to plug in a Votrax speech chip SC-01 or compatible, enabling a phoneme synthesizer capable of unlimited vocabulary, albeit with an expressionless robotic vibe.

Games which boast support for speech through the SC-01 include *Crypt of Madea*, *Rescue Raiders* and *Crimewave*.

Magnavox Odyssey 2 - 'The Voice' expansion module (US-only release, 1983)

After having bought an Odyssey 2 in 1983, you might be feeling a little sorry for yourself. If so, the $100 "The Voice" speech synthesizer would cheer you right up with samples such as "incredible!" and "great!", though being $100 poorer might hurt a little.

The main chip of the voice synthesizer is an SP0256-019 speech chip from General Instrument, as featured in the Intellivision. The enthusiastic built-in phrases took up the 2K of ROM, and more phrases could be added with expansion ROMs, such as an allophone set allowing any word to be created.

Currah μSpeech, commonly referred to as "Microspeech" (1983)

This product was probably the first opportunity UK users would have for meeting programmable voice synthesis, in the form of a simple plug-in cartridge.

The SP0256 variant in this speech module is the AL2, which contains all the English allophones that Odyssey 2 users had to pay extra for. This allowed Currah speech users to curse phonetically with the best of them.

Excitingly for Spectrum owners, it sold enough units for major software houses to include limited support for it, such as Ultimate's *Lunar Jetman*. The game *Booty* even presented a completely different hidden game if it detected the Currah unit.

This model required the UHF output from the Spectrum to be passed through it, allowing it to add the speech to the TV signal.

The 1984 Commodore 64 version

("Voice Messenger" or "Speech 64") cleverly used the Commodore 64's audio input pin to mix the speech into the TV output but has identical output to the Spectrum version. There are no C64 games that support this speech synthesizer, though. This might be because SAM speech already had that market covered, or it might be because it didn't sell as well as the Spectrum version.

Commodore Magic Voice Speech Module (Commodore 64, 1984)

It took Commodore two years after the release of their own machine to produce an official speech module for it, but they got there eventually. The Magic Voice cartridge supported the games *Wizard of Wor* and *Gorf*, both featuring arcade speech, and both released by Commodore. It also supported A Bee C's, a 1983 educational game.

The speech chip doing the heavy lifting is a Toshiba Voice Synthesizer T6721AP, which is unspectacular, merely allowing playback of 235 predefined words from a 16K ROM. I wonder if one of the words was "yawn", in keeping with the levels of excitement generated by the product?

Some inside Commodore even referred to it as *Tragic Voice*, but this did not discourage their marketing department from trumpeting the launch of the TED-based V364 that had built-in speech synthesis using the same chip.

ESS Speech (1984)

In the arcade speech synthesizer section, I mentioned Dr Forrest Mozer, the part-time speech enthusiast with the patented technologies and minicomputers that encoded the speech for Berzerk and numerous Stern pinball machines.

He set up a business with his sons called Electronic Speech Systems to market his technology.

We have them to thank for some iconic video game speeches:

- Impossible Mission (C64, 1984)
- Ghostbusters (C64, 1984)
- Cave of the Word Wizard (C64, 1984)
- Talking Teacher (C64, 1985)
- Kennedy Approach (C64, 1985)
- Desert Fox (C64, 1985)
- Beach Head II (C64, 1985)
- 221b Baker Street (C64, 1986)
- Solo Flight (C64, 1986)
- Big Bird's Hide and Speak (NES, 1990)

AtariVox+ Speech Synthesiser. (AtariAge)

The AtariVox device is a speech synthesizer for the Atari 2600 designed and created by Richard Hutchinson. In addition to period-correct sounding voice synthesis, AtariVox also features an EEPROM that can be used for game save data.

The heart of the unit is the SpeakJet, an amazing chip with phonetic speech, preset sounds and a five-channel music synthesizer. All of these are controlled by the console via a serial interface emulated through the driver software. Speech is constructed from strings of allophones, the individual sounds that make up speech."

Marketing material

(1981) OKI MSM5205 ADPCM

Limited but widely deployed ADPCM decoder marketed for speech.

As seen in: PC Engine CD Add-on, Moon Patrol.

Single-voice sample playback chip. Mono.

The Next Generation

It's not often you see a datasheet warning people off using the chip, but the MSM5205 has one, attempting to upsell to the next generation.

> "For a new circuit design, it is recommended to use the MSM6585 as described later. The MSM5205 has a 10-bit DA converter and does not have a built-in low-pass filter.
>
> On the other hand, the MSM6585 has a 12-bit DA converter and includes a -40 dB/octave low-pass filter.
>
> The sampling frequency can also be selected up to 32 KHz. Therefore, the MSM6585 can realize a high-quality voice."
> **MSM5205 datasheet**

So, before we all rush off to buy an MSM6585 (only used in a few pinball machines), let's look beyond the negativity at the MSM5205 itself.

It's a modest, monophonic ADPCM decoding chip used in a wide variety of mid-'80s arcade games, such as *Moon Patrol*, *Kung-Fu Master*, *Lode Runner*, and *Gladiator* aka *Great Gurianos*.

Speech, not speech

The marketing team behind this chip knew their stuff — it was marketed as an "ADPCM SPEECH SYNTHESIS LSI" rather than a sample playback chip.

Speech is mostly what it was used for, such as short phrases and the occasional maniacal laugh. However, later in its life, it was also pressed into service for SFX, and even to provide additional percussion or instruments as part of a multi-chip setup.

For example, *Gemini Wing* uses it for bass/snare, and *Silk Worm* has a rhythm guitar, both accompanying a Yamaha YM3526 OPL1.

MSM5205 accepts 3-bit or 4-bit ADPCM @ 4, 6 or 8 KHz.

Looking at the datasheet, this data isn't coming directly from a ROM, so it must be fed with data by the CPU.

They just kept coming...

Oki were correct that the MSM6585 was a better chip, though enough companies ignored the upsell to give this silicon an incredible 13-year lifespan in coin-ops.

This modest silicon even put in a cameo appearance on the TurboGrafx-CD/CD-ROM² add-on (1988) which allowed the PC Engine to use CD-ROM games, the first console to do so.

(1981) NEC μPD1771C

Raucous and noisy one-voice SSG often programmed as a two-voice.

As seen in: Epoch Super Cassette Vision (SCV), NEC APC, Grandstand Firefox F-7 Handheld.

Single-voice SSG with wavetable. Mono.

Did you ever…?

If you ever flew single-handedly in a Grandstand *Firefox F-7* handheld game, then you've heard this chip, though probably through tinny speakers.

If you've ever used an NEC APC, a computer so business-like that one of the clocks is actually called "The Business Machine Clock", you've heard this chip, albeit not at its best.

And if, by some chance, you were a Japanese consumer with a phobia of Nintendo and Sega but a love of cartridge games, you'd have bought the follow-up to the 1982 market leader, the Epoch Cassette Vision. This machine for the ages, the Epoch *Super* Cassette Vision, sported this sound chip in all its 1-channel wonder, with aggressive tones, noise and PCM. "Cassette" in Japan means "Cartridge", not magnetic audio tape.

The business of sound

As a console to compete with the Famicom and Master System, the available games list is on the short side, with the main stars being ports of *Boulder Dash* and *Miner 2049er*.

But it's not often you come across such a flexible one-voice chip as this one, so let's go back a step and look under the hood with the manual for the NEC APC, which has a slightly different variant of the same chip:

> "The sound control system is supported by the NEC µPD1771C-006, which drives a loudspeaker through a 4-watt audio amplifier. This LSI generates audio signals and programmable music.
>
> Through music programming, the sound control system can generate tones ranging over two octaves in frequency, at specified note lengths, intensities, and tempos."

First thought: "only two octaves"? That's pretty small.

Secondly, "specified note lengths"? Tempos? Well, yes. The chip isn't doing that, the NEC itself is, including such features as "sharp attack", "soft attack" and even "moderately emphatic rhythm".

Don't interrupt! Oh, go on then…

That NEC write-up doesn't sound a lot like what's going on in the SCV, though. Listening to the game soundtracks, you can hear that they're making great use of "a complex noise and tone internal interrupt system" and that many games are effectively using two voices (though there is never musical two-voice polyphony).

That is, there are two (sometimes more) sounds at once, interleaved together so there's only one actual sound playing at a time, sharing the output on a time-slice basis.

Interleaving the sounds is a technique also seen on the DAC for the Sega Mega Drive, and as with that machine, it gives the sound a rough, choppy tone.

The games drive the chip hard, doing echoes and dynamic envelope generation in code.

It's not a million miles away from the raucous two-voice TIA sound (which, I guess, is the point), sounding like a stereotypical console.

The reason for a wide variety of sounds coming from this chip is that it also supports wavetable playback, but it's up to the cartridge to provide the sounds to fill its huge 64 bytes of RAM with playable waves!

Emulation development documentation from Japan's Toshiya Takeda hints that the waveform selection into the ROM is a three-bit register (i.e. eight sounds), with five bits for specifying the frequency. However, it also has a noise generator and a square wave tone generator, and the monophonic voice can be switched between them. There are hints about other features, but a considerable degree of uncertainty about what to make of them and how to implement them:

> "The noise command seems to be 10 bytes. The meanings of the parameters are as follows. $01, $80 / $e0, WN wavelength, WN volume, square wave wavelength, square wave attenuation ,?, Square wave volume,?,? What mode is $80 / $e0 specified? WN is a white noise-like waveform. Square wave attenuation attenuates the volume within one pulse. Is it more like periodic noise than white noise? The waveform itself seems to change dynamically, but for the time being, a sample of one waveform is used as an array. Let's have it so that it can be played easily."

It's all so obvious now!

This wasn't even the weirdest thing that happened that day...

Commodore 64 Tape Loaders

On the Commodore 64, loaders first went nova, then went interactive!

It's a cliché that loading all 8-bit games from cassette tape used to take forever. It's certainly true on the Commodore 64 that the ROM routines for load/save are terrible.

> "I looked at the ROM load and save routines and was just flabbergasted at just how poor they were!
>
> No wonder loading was slow; it effectively stored more than four times the data it needed to! Every byte stored required 20 pulses – two for each bit sent, plus a couple of parity bits and a couple of bits to mark that this byte was the last byte (!).
>
> This data was then stored again to verify the load!"
> **Paul Hughes, Freeloader creator**

The criticism is also unfair. Modern games might easily keep you occupied for an hour while they download updates, and a game like GTA V can take a while to load.

Anyway, fast loading solutions did not take long to appear on the Commodore 64, reducing loading time to seconds rather than minutes. I have a vivid memory of being shocked at *Son of Blagger* and *Revenge of the Mutant Camels* loading in under 40 seconds, both using a speedy loader programmed by Kingsoft.

That loader was "blank screen". While the ZX Spectrum had always had flashy-border loaders, the first time Commodore 64 users saw something similar was in Pavloda. This was seen on games such as *Cavelon* and *The Way of the Exploding Fist*.

The latter had no loading music but did surprise everyone with an iconic Bruce Lee scream during loading!

Loaders go Nova

It was the late, legendary Paul Woakes of Novagen who in 1984 created the first musical loader on the Commodore 64.

With loaders long having been silent affairs, this one broke the mould by allowing you to hear every byte of the loading, along with "whooshes" and "whees" periodically as the pattern of data created accidental sound effects.

That atmosphere suited the game the loader was attached to, which was his amazing 3D game *Encounter*. But it was always going to grate on the nerves for other games, and so a music engine was tagged on.

The engine is only suitable for rudimentary tunes. This is partly because the sound can't be changed after the tune has started, and partly because the composition must be entered in a BASIC editing environment. The timing of the notes is also occasionally suspect on playback.

Nova Love

The most heard version of Novaload was in the US Gold games of the time such as *Tapper*, accompanied by grindy US-centric tune medleys arranged by Gary Sabin. They were musically accurate, but over-use made them increasingly annoying.

Ocean's first outing with Novaload for games such as *High Noon* offered "A Policeman's Lot" from the Gilbert and Sullivan opera "Pirates of Penzance".

Also standing out are Mark Cooksey's use of Trans-X's pop hit "Living on Video" to accompany Elite Systems titles, and Jeff Minter's ambitious but slightly dodgy part-cover of the Genesis classic "Los Endos" for *Ancipital*. They were both lucky to get away with that! Jeff also hacked a small light show into *Batalyx*, with flashing squares following the tape loading noise!

As time wore on, games such as *Thing on a Spring* and *Monty on the Run* moved to Novaload 2, using hi-res loading screens and no music.

Doing it the Gal Way

Ocean's Bill Barna decided he could improve on Novaload, and he did that by removing most of the guts put there by Paul Woakes and keeping just the tape loading routines. In some games (*Hunchback II*, *Kong Strikes Back*), the loader displayed a hi-res screen loading image and had no music. But there was one loader that Bill hacked to include interrupt-driven music that was a massive improvement on Novaload's.

That loader was *Daley Thompson's Decathlon,* and it used the talents of one Mr. Martin Galway, fresh from school and with an abiding love for the music from Sega's arcade game *Super Locomotive.*

He had already recorded it to tape from the arcade machine and programmed it on a BBC micro. Now, sitting in Ocean's offices (because he didn't have a Commodore 64 yet!), he typed in the notes he knew so well: the notes to Yellow Magic Orchestra's "Rydeen".

A straw poll of C64 users today reveals that Ocean and Martin had created a life-changing event. But did anyone realise that at the time?

Personal Computer Games magazine (PCG) only gave the game 6/10 for sound, but the main review mentions "a superbly colourful and musical turbo loading sequence". Your Commodore magazine (December 1984) doesn't mention the music at all, and nor does Commodore User (November 1984).

So, overall, the music made a much bigger impression on the fans than the reviewers.

In the same issue, PCG had a news article about turbo loaders, in which it was announced that:

> "Meanwhile, Commodore have announced that their future cassettes will use the Nova Load [sic] turbo system first seen on Ocean's game Daley's Decathlon. This system allows music and moving screen displays while the game is loading. Ocean say it is now set to become the industry standard."

There is a lot wrong with this statement. *Daley's* didn't use most of the Novaload code Commodore was adopting, Novaload was first seen in *Encounter*, and Ocean were in a gradual process of evolving Novaload into the iconic Ocean Loader, first seen in Roland's Rat Race, one of Martin's first pieces, created before his driver acquired Pulse Width Modulation abilities in Cyclone/Helikopter Jagd.

Fun fact: there actually was a short-lived attempt to start a tape loader in November 1984 in the same driver as Kong Strikes Back (before Martin properly joined Ocean in January 1985). Some Ocean waves and a start-up jingle are all that remain.

The piece now known as "Ocean Loader 1" was never meant to be called "Ocean Loader" at all. The original idea was for each Ocean game to have its own loader.

The tune we know as "Ocean Loader 1" was the "*Hyper Sports* Loader". Ocean Loader 2 was the "*Comic Bakery* Loader". However, while some tunes received a memorable Galway treatment (*Rambo First Blood 2* being a loader as iconic as *Hyper Sports*), most had to share their loader with other games.

Whatever it's called, the *Hyper Sports* loader blew minds and put cassette loading music on the map. Julian Rignall of Zzap!64 called it "*a real megatune*" and it was highly praised by Jeff Minter in his "Nature of the Beast" newsletter, which is where the discerning C64 fan got their industry gossip in those days.

There is also an Ocean Loader III from Martin that was never finished as he left the company shortly after starting it. After the customary opening, it bursts into an upbeat and uplifting three-chord sequence with the melody played on a filtered bass.

Paul Hughes' arrival at Ocean followed a bit later by Jonathan Dunn's meant an increasingly powerful "Freeloader" system and more memorable loading experiences for a flood of games, compilations, and re-releases, with music from Jon (and latterly, Matt Cannon) no longer based on Martin's venerable driver.

Back in 1985, Martin's original *Hyper Sports* loader had raised the bar for both music and loading experiences.

Ocean's Bill Barna was also put to good use implementing the *Frankie Goes To Hollywood* fast loader with Fred Gray's stunning filter-heavy cover of "Relax", for example.

Imposing Sanxion

The next giant step forward combined two major talents: Rob Hubbard and John Twiddy for the *Sanxion* loader. That tune, later named "Thalamusik", probably to keep Zzap!64 happy, was the star of the first loader to use John's new Cyberload system. By that time gamers were used to turbo loaders, but *Sanxion* was special, with its scrolling message and mysterious glowing Thalamus logo.

It was also Cyberload that played host to classic loading tunes from two more of the big names: Ben Daglish and Matt Gray. John Twiddy's System 3 connection guaranteed that, giving us loaders for *The Last Ninja* (Ben, with Anthony Lees) and its sequel (music by Matt). Ben had been in the industry since 1984's *Percy the Potty Pigeon*, but *The Last Ninja* "Wastelands Loader" is widely regarded as one of his masterworks.

Matt's loading music debut had been with the *Que-Dex* loading music (yes, that's the correct punctuation!), originally written in off-the-shelf composing package *Electrosound*. He also stepped in when Rob Hubbard left for the US in 1988 to provide Rob-Hubbard-impersonating music for the *Bangkok Knights* loader. Many fans prefer the loader to Rob's in-game music, especially fans whose filter made that tune sound terrible.

Mix-A-Lot

Backtracking a little, it took until 1987's *Delta* for Rob Hubbard, with heavy encouragement from Zzap!64 alumnus and Thalamus producer Gary Liddon, to create "Mix-E-Load": essentially a create-your-own-loader-tune which allowed you to mix and match a selection of basses, accompaniments and leads. Jeroen Tel's *Hawkeye* Mix-E-Load built on this idea magnificently the next year.

Game within a game within a game within a…

There's one inception-level advance in turbo loaders to go, and that's the game before the game.

Mastertronic's *Invade-a-load* was a clever *Space Invaders*-based minigame that also functioned as a turbo loader. Written by Richard Aplin, it used Rob Hubbard's pre-existing main theme from *One Man and his Droid*. Not that Rob (or indeed, Taito!) knew of this, but that's the '80s software industry for you!

Dodgy IP use aside, *Invade-a-load* made quite an impression on users, but not on history, as Wikipedia explains:

> "In 1995, Yoichi Hayashi of Namco Ltd. invented a variant of this technique for use with optical disc-based platforms such as PlayStation and applied for a patent. U.S. Patent 5,718,632 was granted in February 1998 and assigned to Namco despite the Invade-a-Load prior art."

It seems Mastertronic was not around to stop them.

There was also *Snakeload*:

> "SNAKELOAD V5.0 (C) STEVE SNAKE, KML 1987. THIS IS NOT SUPPOSED TO BE PROTECTED
>
> SO DON'T THINK YOU ARE GOOD FOR CRACKING IT !!!KYLIE RULES!"

This loader replaced *Space Invaders* with the classic *Snake* game that people think Nokia invented, because people are dumb. Not you, though. You're great. Did you do something with your hair? Looking good!

More tech info about tape loaders? Here you are:

https://www.luigidifraia.com/c64/docs/tapeloaders.html

(1982) Commodore MOS6581/6582/8580 SID

Legendary and flexible PSG with an almost religious following… and chart success!

As seen in: Commodore 64 and family.

3 voices (choice of pulse/square/triangle/sawtooth/noise/mixed waveforms of varying usability).Full ADSR envelopes, ring-mod, hard sync, LPF, BPF & HPF analog filters and audio in (use carefully!).Noise channel has a 23-bit LFSR with feedback taps at 17 and 22.

Bob and Rob, pioneers

If the AY-3-891x is Coke and the SN76489 is Pepsi, then the MOS 6581 SID is Inca Kola. It's a world apart from the competition, thoroughly idiosyncratic, and plays by its own rules.

The reason for that was Bob Yannes.

In 1996, he was interviewed by Andreas Varga of the High Voltage SID Collection (HVSC), the primary archive of music programmed for the chip. Quotes from him are taken from that interview.

Bob was inspired by a desire to improve musical standards in an industry where sonic mediocrity was endemic. Rob Hubbard entered game music for the same reason, perceiving a lack of basic skill and musicality in the marketplace.

> "I thought the sound chips on the market (including those in the Atari computers) were primitive and obviously had been designed by people who knew nothing about music."

It's true that earlier chips tended to be "good enough", and had generally been used more for sound effects, despite some being musically capable.

> "I was an electronic music hobbyist before I started working for MOS Technology (one of Commodore's chip divisions at the time) and before I knew anything at all about VLSI chip design. One of the reasons I was hired was my knowledge of music synthesis was deemed valuable for future MOS/Commodore products."

Bob was on a journey from electronic music hobbyist to a synth technology god.

That journey led through the VIC and then the SID, a chip that Bob hoped would be good enough to tempt synthesizer manufacturers to incorporate it into their hardware.

Commodore ex-employee Andy Finkel remembers showing a prototype SID synthesizer at one CES, that sported three SID chips (and a fourth in the Commodore 64 that was running the system). Each SID chip powered one extremely powerful, three-oscillator, single-channel voice.

Commodore models and their SID chips

1982	C64 (Breadbin)	6581
1982	CBM-II	6581
1982	Max/Unimax	6581
1983	SX-64	6581
1984	Educator 64	6581
1985	C128	6581
1985	C128D	6581
1986	C64C	8580
1987	Aldi 64	8580
1988	C128DCR	8580
1990	C64G	8580
1990	C64GS	8580
1993	C65	2 x 8580

Bob Yannes left Commodore in 1982, just as the Commodore 64's commercial life began. His desire to create professional studio synthesizers was realised when he co-founded Ensoniq shortly after his exit from the home computer market.

During the lifetime of the C64, very few SID-based products were released. In 1989 Innovation Computer developed the Innovation Sound Standard, an IBM PC compatible sound card with a SID chip and a game port.

The success of the Commodore 64 line meant that virtually all the SID chips produced by Commodore were reserved.

SID-based music hardware had to wait for 1997's "SidStation", a limited-edition synthesizer released by independent company Elektron who also had the foresight to buy most of the remaining SID chips, as production lines for the chip had shut down.

Another pioneer's story began in 1983.

Rob Hubbard was a gigging band member and electronic music hobbyist with studio experience. The advanced features of the SID chip drew him in.

The idea of a career programming music on it was as unlikely to him as it would have been to anyone else in 1982. However, the buzz around home micros meant that there was a sense of "computers are the future".

If the C64 had an AY-3-891x or SN76489 chip, would Rob have made the switch to computers? Rob thinks so:

> "Yes, because the wider issue was really about learning about the computers and getting into programming. But things might have turned out quite different without the game music side of things."

The SID is fun. It is an adventure to program, though it is also limited and frustrating. It is full of complexity, twists, and turns, but also buggy.

It's a proper, programmable synthesizer, but without a user-interface. Initially, it was also underused by most games.

It was that sense of wasted possibility and the opportunity to improve on the status quo that was the driving force behind Rob's game music career.

Having been advised that his games weren't good, but his music was, he made the mental leap from "programmer" to "musician".

If Rob had felt there was no real reason to improve game music, he might have just used his programming skills for a respectable business programming career.

A crowd was gathering...

Future star composers such as Martin Galway, David Whittaker, Ben Daglish, Fred Gray, Graham Jarvis, and Russell Lieblich were already on the scene and on the SID producing impressive sounds before Rob made the leap to "game musician".

For instance, Fred Gray's *Shadowfire*, Martin Galway's *Kong Strikes Back* and Ben Daglish's *Loco* soundtracks were all released prior to Rob's first two released games, *Confuzion* and *Action Biker*.

However, Ben, David and Fred are all on the record that hearing Rob's stuff for the first time inspired them to new heights of music and driver programming.

This friendly competition intensified as popular magazines like Zzap!64 started to treat the composers like rock stars and undoubtedly contributed to the massive strides forward in game music on all platforms between 1985 and 1988.

It all started here

What was so exciting about the SID? It offered new waveforms and sound techniques, redefining what home micros could do with sound.

Techniques and sounds originally developed for the SID chip were soon backported to simpler chips as C64 composers adventured to other platforms.

A signature technique that started to appear was toggled chords ("arpeggios"). The composer would play the individual notes in a chord so quickly one after the other with a single voice, giving the impression of a single chord.

It became an iconic chipmusic sound. In fact, it became so iconic that it was later perceived as almost childish, meaning that Rob Hubbard had to remove it from his Electronic Arts music driver when he started writing C64 music for primarily American audiences. These chords were sometimes replaced with a sampled organ part.

More than a square wave

The common square wave (with its 50/50 on/off ratio) became a pulse wave. The SID allows the percentage of "on" to be defined and even changed while the sound is playing.

Changing the pulse "width" while the sound is playing is a signature sound of the SID. While the NES allows it to be changed, it only offers four widths, two of which sound the same.

More powerfully, sawtooth and triangle waveforms were implemented, along with the traditional noise generator, doubling as a pseudo-random number generator.

Each waveform can be used on each voice, with powerful ADSR to control the shape of the sound.

The SID also allows the waveforms to be mixed, though deciding which results are usable is up to the programmer and can even vary between chip models.

These mixed waveforms appeared in a few showcase pieces, such as Martin Galway's in-game music for Sensible Software's *Parallax*, nicknamed "Parallax Stroll".

Like pitch, the waveform of a sound can also be switched without restarting a note. This allows programmers to cleverly trick the ear. For example, composers often use a short burst of the noise waveform at the beginning of a note to emulate a hi-hat. This means that technically the note that followed it is late, but the ear is easily fooled.

SID also features a powerful but unreliable (and very difficult to emulate) analog filter that can do low-pass, high-pass, and band-pass filtering, all at the same time!

Luxury features include filter resonance ("makes things go sweeeep!") and the ability for voices to modulate other voices with "Ring Modulation" and "Hard Sync", which is heard in tunes such as *The Last Ninja*® ("Wilderness", a grindy tune using hard sync) and *Spellbound* (ring modulation for bell sounds and the "tearing the SID chip apart" sound).

The Commodore 64 would have been a success without these esoteric features, so putting them in seems like (welcome) self-indulgence on the part of Bob Yannes.

Due to a short schedule and chip design tool limits, Bob's original plan for SID was revised from 32 voices, to 8, then to just three. The music produced would have sounded very different if composers had not had to fight a three-voice limitation.

Buggle Bobble

The more complex a system, the more likely it is to have bugs and imperfections, and the SID is no exception.

The filters are great, but massively variable between individual computers. This means occasionally inaudible sounds on filter-heavy tunes such as Rob Hubbard's *Light Force* or *Shockway Rider*, or Martin Galway's *Miami Vice*.

The envelope generators are sometimes inconsistent, leading to sync and timing problems in the music colloquially known as "the rubber band effect" or "the school band effect".

The filters are an analogue component in a digital sound-chip, their efficacy and effects even altering as the chip warms up, though the main cause of filter differences between chips is the massive quality/manufacturing variance in capacitors.

The biggest bug is with the volume register on the 6581, which makes an audible "click" whenever it is changed.

> "It's not actually a click like most people would think. There is a DC offset at zero volume which is amplified when the volume is increased. The centre point of the output waveform moves up or down relative to the volume level. You literally just have to stream 4-bit samples through the volume register at whatever sample rate you want."
>
> **Mike Clarke, creator of "InSIDious" VST (from Impact Soundworks)**

By driving it at speed with interrupts, sample playback is possible with low CPU usage. This bug opened the door for software such as *Digidrums* and *Microrhythm*, as well as the first sampled percussion in a C64 game (*Arkanoid*, by Martin Galway), and then later the first use of tuned samples in a home micro soundtrack (*Arcade Classics*, by Rob Hubbard).

This wasn't the only sample playback method that was developed for the machine. A "pulse-width" technique was occasionally used.

According to demo-scener and programming pioneer Pex "Mahoney" Tufvesson, it was rough:

> "There is the "pulse-width"-technique for playing samples, but that has even worse carrier noise, and would not be able to get any higher frequencies than the highest pitch the SID chip can produce, which is around 4kHz".

He also mentions another technique developed more recently:

> "There is the 'one voice, test-bit, triangle-wave, sample-and-hold by just briefly enabling waveform"-technique invented by Otto "SounDemoN" Järvinen in 2008. It has been used properly in 16kHz, but with very low output amplitude and around 5-6 bits resolution."

However, being the demo-scener's demo-scener, Pex found his own way of playing 8-bit audio at near-CD-quality and introduced it in a demo called Musik: Run/Stop in 2014.

He helpfully wrote a white paper about his technique, and it's that document I've been quoting here.

The technique is too complicated to explain fully here because it relies on the intersection of three quirks of the SID, but please check out his paper. http://qr.c64audio.com/video.html?video=-17

Better than the arcades

A very odd thing happened to some games on their way from the arcades to the C64.

A combination of enthusiasm and lax creative control from arcade game manufacturers (usually located far away in the USA and Japan) resulted in C64 soundtracks with extra music (*Mikie*, *Arkanoid*, *Terra Cresta*, *Space Harrier*, *Ghouls 'n Ghosts*, *Turbo OutRun*), massively upgraded music (*Commando*), or music completely different from the arcade machine (*Ghosts 'n Goblins*, *Paperboy*).

The Golden Years

The number of SID composers and releases rocketed, thanks to trackers, demo parties, disk swapping, the early online service Compunet, and magazines that covered the whole "scene".

> "I think the C64, and its SID chip, tore up the rule book for audio designers, sound effects and musicians whose creativity & originality whisked them away from back bedrooms into the new full-blown game studios.
>
> An army of creatives from coders to musicians found their niche at an affordable price & time that paved the way for the gaming future..."
> **George Bray, industry veteran**

The C64 was also notable for hosting much longer and musically ambitious tunes than other platforms. For example, the C64's version of *Tetris* had an amazing 25-minute ambient soundtrack by Wally Beben rather than the unashamedly plinky music of the Atari-based versions.

Works such as *Sanxion Loader*, *Wizball*, *The Last Ninja* or *Knuckle Busters* still sound unique today, and many of the compositions were strong enough to be remixed and adapted into real-world genres as diverse as dance, country and western and classical.

SID - the historical chip

SID tunes occupy a unique place in sonic history. The underlying composition is sophisticated enough to be exciting, and the sounds can be complex enough to reference a wider musical context (for instance, Rob Hubbard's solo violin in *Master of Magic*).

However, the listening experience of SID isn't complete and fully formed. It's impressionistic.

Rob's solo violin is, for example, a powerful impression of a violin, but it requires the listener to interpret it and draw that connection, and this is an involuntary creative process.

Studio tracks generally don't require the listener to interpret them because they're musically complete, but SIDs require work to fully enjoy. Sometimes, as in pieces like *Monty on the Run*, the listener must sometimes imply their own chords because the available musical information is ambiguous!

The later move to sampling on platforms such as the Commodore Amiga removed a lot of the need for interpretation because the sounds were "real".

However, those sounds were also low-resolution, and even some iconic Amiga composers such as Allister Brimble now feel that their SID work has aged better, especially when they return to their Amiga work and try to remix it themselves!

Regrets, Bob had a few

Interviewed in 1996, Bob was pleased people had liked the chip, but was mostly concerned with its limitations and problems, especially as he had not personally experienced the music being created with it:

> "The design/prototype/debug/production schedule of the SID chip, VIC II chip and Commodore 64 were incredibly tight (some would say impossibly tight) - we did things faster than Commodore had ever done before and were never able to repeat after!
>
> If I'd had more time, I would have developed a proper MOS op-amp which would have eliminated the signal leakage which occurred when the volume of the voice was supposed to be zero.
>
> This led to poor signal-to-noise ratio, although it could be dealt with by stopping the oscillator.
>
> It would also have greatly improved the filter, particularly in achieving high resonance.
>
> I originally planned to have an exponential look-up table to provide a direct translation for the equal-tempered scale, but it took up too much silicon and it was easy enough to do in software anyway."

SID expert, journalist, and composer Andrew Fisher notes that the chip experienced multiple changes in its long lifespan:

> "The SID story did not end with the original 6581. Commodore, in seeking to reduce even further the costs of producing each machine, created the C64C. This housed a revised 8580 SID chip using new manufacturing processes, and in the process of remaking/remixing the chip, there was an unfortunate side-effect.

In 'cleaning' up the stray voltages and signal leakage, the 'hack' that allowed the playback of samples via the volume register was badly affected. The result is that sampled sounds that are loud and clear on the 6581 silicon are barely audible on the later 8580 without using a 'DigiBoost' coding trick or hardware boost.

Many SID composers swear by (or at) one revision of the chip, targeting their music and sounds for that model. Fortunately, emulation is sufficiently mature to cope with the differences.

Commodore's lack of confidence in their technology even went as far as prompting them to remove the details on the Sync and Ring Mod registers from the C64C User Manual - but they remained on the chip and can generate some pleasing effects."

Hardware guru Tommi Lempinen points out that while there was a later 6582A SID, it's identical to the 8580 (down to the die) and the name change was for external marketing of the chip.

Influential, rather than famous

Millions of C64s were sold, and the platform played host to hundreds of quality composers, and tens of thousands of SIDs, with activity ongoing. Techniques honed on the C64 continued to influence chipmusic on other platforms, (in the West, at least).

However, the lack of culturally significant brand IP on the platform meant that the C64 sound didn't make much of a cultural impression.

Chipmusic in the popular mind is now mostly synonymous with Nintendo franchises such as Tetris, The Legend of Zelda and Super Mario Bros, and the sound of the NES and Game Boy.

However, composers who cut their teeth on the C64 went on to illustrious careers, providing the soundtrack to the lives of countless gamers right to the present day.

And the chart success? Zombie Nation & Timbaland (both legally controversial), Scooter, Console and others all charted with the sound of SID, almost always from the SidStation.

(1982) NEC μPD7751

Functional sample playback that was a precursor to better things from Sega.

As seen in: Pre-System 16, System 16A

Monophonic ADPCM decoder capable of 14-20 Kbps playback. Sampling clock 4, 5 or 6 MHz. The chip can address an arbitrary number of 16K ROM chips, containing eight "messages" (samples) in each one.

Before Sega got players ready for the Fantasy Zone with their all-talking, all-screeching, all-drumming SegaPCM chip (*Out Run*, *After Burner*), they still needed ways to tell players vocally to get ready, or that they sucked.

Stepping up for vocal duties on Sega's "Pre-System 16" and "System 16A" boards was this chip, the catchily named μPD7751. It was originally designed for high quality telephony and speech encoding but, as is common, was repurposed.

The chip boasts "combined voice and background music capability", which means the speech-optimised encoding/decoding processes don't degrade musical content too much.

For its deployments, it was the only sample/speech chip on the board. As it's monophonic, the games had to be carefully designed so that the system is never faced with the job of playing two samples at once. Because it was often paired with a YM2151 and its "effects channel", this chip was only ever needed for basic monophonic playback.

Games with this silicon on board include *Monster Bash*, *Alien Syndrome*, *Quartet* and *Time Scanner*.

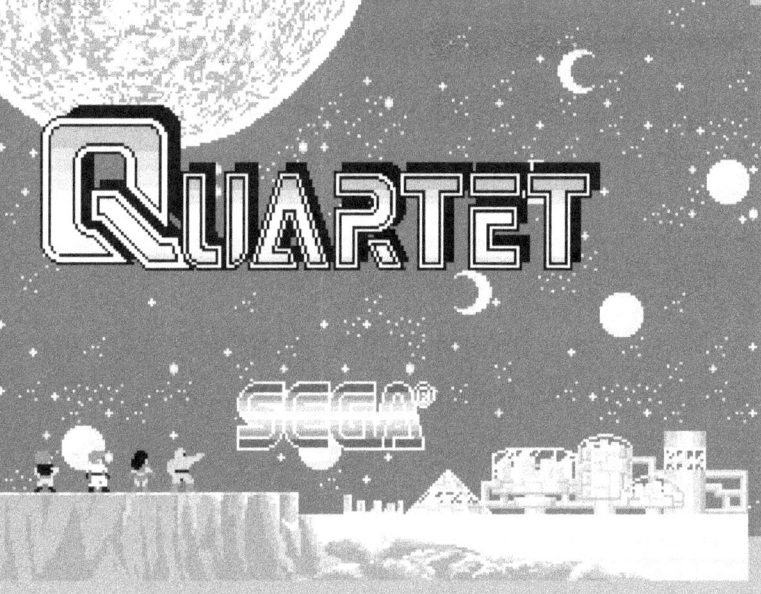

(1983) OKI MSM5232RS

Unusual but delightful organ-pipe obsessed SSG and noise generator.

As seen in: Bull Fighter, Metal Soldier.

8-voice SSG. Mono.

OKI MSM chips tend to be ADPCM-playback or speech-synthesis chips, but this chip breaks that rule.

It's an interesting and quirky synthesizer chip found mostly in Taito games from 1984 - 1986, though many of the titles aren't familiar to gamers of the time.

This chip also appeared in the Korg Poly-800 keyboard.

We've come across the situation before where a sound-chip gives voices different ranges to try and improve the pitch accuracy. The VIC-20 does it, for instance, by having bass, alto, and treble voices - each one capable of a certain range of notes.

When I go high, you go low

This chip has a unique architecture to produce its sound. It's capable of a 7-octave range in total, but it forces you to group voices into two sets of four, and then designate a four-octave range for that group.

For instance, four of the voices might be "bass" and four of them "treble". This is like specifying "I'll have four bass singers and four treble singers, please!".

Each of the two groups of four is treated as one distinct unit and the voices within it share the same volume levels and envelope shape. The choice of envelopes is between "lasting sound" (like an organ) or "damping sound" (like a piano).

Behold, my mighty pipe length

Rather than using "bass", "tenor", "alto" and "treble" for its four octave settings, the chip uses feet - as in, the measurement - 2 feet, 4 feet, 8 feet and 16 feet, where bass is 16 feet, and treble is 2 feet. This refers to a pipe length.

Think of a didgeridoo vs a flute: a big pipe vs a small pipe. The big pipe has a much lower sound and isn't capable of the same high sounds the flute is. But there is probably a little overlap. Even with a pipe of the same radius, a longer pipe has a lower sound.

If you look at a church organ up close, you can see 2', 4', 8' and 16' written on "stop tabs" (pullable knobs), that control the amount of wind going to each pipe.

As an aside, church organs often had stop tabs with the name of instruments on them: proof that even in centuries gone by, musicians were suckers for factory patches and in-built sounds.

Mmm, convenient

Frequency-to-note calculation is rarely straightforward, and OKI handily included a table to allow the programmer to look up the frequency to generate for a given musical note on each "pipe" setting.

While the chip claims to be only 8 voices, the noise signal coming from the chip is on a separate pin - making it essentially another voice.

> "This is the noise sound source output terminal that always outputs noise generated by the incorporated pseudo random pulse generation circuit."
> **Datasheet**

Down the arcade

The MSM5232RS was often accompanied in the games by one or two AY-3-8910s or YM2149F chips.

Sometimes the AY/YM SSGs add depth (the two synthesis methods blend very well), rhythm (the MSM is not good at short, stabby sounds but the AY is excellent at them), echo or delay (by doubling a voice).

Sometimes it handles all the music and lets the PSG handle the sound effects.

Overall, this chip improves the musical experience markedly, with composers exploiting the extra voices for richness and depth.

Games, games, games

Alpha Denshi's Bull Fighter, surprisingly, is an Ice Hockey game, though no one seems to have told the composers who produced a spirited and catchy rhumba more suited to the chip's home organ origins than the arcade. The game throws in some speech as well from a Hitachi chip.

Metal Soldier by Taito boasts a title tune that's a particularly good melding of the types of synthesis and has a lot going on. It's not often you get a unique sound in the arcades, but this is well worth a listen.

Taito's The Fairy Land Story, a forerunner to Bubble Bobble, uses the organ to beautifully pad out the music with swelling chords and a lush sound, adding a delightful "home-made" atmosphere.

MAME emulation of this chip is imperfect: for instance, the sounds tend to die away too slowly, but at least this makes them easy to spot.

The octave doubling on all the notes contributes a real church organ vibe which adds drama, if only the hammy "Toccata and Fugue" kind of drama, rather than the "Finish Him!" kind of drama.

Not all games used this chip to its limits: Buggy Challenge, also by Taito, wastes it on a dull fanfare and underwhelming music.

The Fairy Land Story.
For those reading in black and white, the dragon is the blue thing!

(1983) SNK Wave

A cheap rip-off of Namco's WSG with limited deployment.

As seen in: Marvin's Maze, Vanguard II.

Single-voice wavetable. Mono.

WSG Wannabe

Every chip has cheap imitators, and this one is a cheap imitation of the legendary Namco WSG (the wavetable force behind Pac-Man). It's fitted to what's been called SNK's "Triple Z80 Board" (two Z80s for the game, one for the sound).

It's a single-voice wavetable chip running at 8 MHz. The wavetable consists of only 8 steps, each with 3-bits of resolution (so, four amplitude levels are possible) which is inferior both in length and depth to Namco's device.

An oddity is that the wave isn't looped but is played forward and backward continually through a crunchy 4-bit DAC.

12-bits of frequency register is surprisingly precise considering the under-specced wavetable section, but the same as the twin AY-3-8910s it is deployed with for the three games that use it.

Thanks to these capable if unspectacular music chips, the SNK Wave doesn't have to do any musical heavy lifting itself.

Listening to the games, it's difficult to know what these chips were used for that couldn't have been done by one of the AY chips, though some of Vanguard II's sounds have a slightly ring-mod feel to them, and Mad Crasher has a very annoying crunchy background noise.

(1983) NES APU – Ricoh 2A03 (NTSC)/2A07 (PAL)

Four-voice PSG and a fifth voice for samples, with a sound as quirky as its design.

As seen in: Famicom, NES

Five voices, consisting of two pulse voices (with four selectable duty cycles), one triangle, one noise with 15-bit linear feedback shift register, and a sample channel supporting DPCM-compressed 8-bit audio.

The voice of chiptune

To a sizable chunk of the world, the Nintendo Entertainment System/Famicom (and its little brother the Game Boy) is chiptune. This is in no small part because of an Italian plumber, some coloured blocks, and a Robin Hood lookalike.

While the Japanese market got to experience the Famicom from 1983, Westerners only discovered its delights in 1985.

The tech behind the tunes

The CPU that powered the Famicom is based on a probably-modified-to-avoid-patent-royalties 6502, that also contained the APU sound generator - the secret behind World 1-1 and other iconic gaming moments.

The APU has five channels in total: two pulse wave generators, a triangle wave, noise, and a sample channel, in this case a delta modulation channel for playing DPCM samples.

It's a little odd to find such a sophisticated and quirky piece of circuitry on an otherwise cut-price CPU chip, but it certainly played its part in the NES's extraordinarily long lifespan.

The NES arrived in Western shops looking very different from its Japanese sibling because Nintendo wanted it to look like a toy rather than a computer.

The specifications didn't change too much from the original Famicom. The main audio change was the removal of the audio link between the cartridge port and the console. This is a Famicom feature used by additional sound chips embedded in some later cartridges such as *Castlevania 3*. Without mitigation, this would result in missing audio on the NES version of an enhanced Famicom game.

Sonically, the square wave has four widths. The widths available are 12.5%, 25%, 50%, or 75%, with 50% being the pure hollow-sounding square wave. While having only four settings would appear to be limiting, there is a "sweep unit" which can be programmed to "sweep" the pulse width setting up or down across time. It's still more limiting than SID's ability to specify any width. It's puzzling why they have two settings (25% and 75%) that sound the same, when they could have had 66%.

It's also odd that there's only one sound-shaping envelope option in the hardware, which is a sawtooth envelope. It can be applied to both pulse and noise channels or bypassed in favour of software envelopes.

Triangle: the bassy waveform

The triangle wave stands out as an oddity in this setup. The output can't be shaped, and it doesn't even have a volume control, except by abusing the DPCM channel, which *Super Mario Bros* does between levels. It's also pitched an octave lower than the pulse wave channels, and thus used so frequently for basslines that it almost defines the NES sound (*Super Mario Bros*' World 1-1 being the prime example).

Dynamic Pulse Crunch Modulation

The sample channel was forward-thinking, but not often used. DPCM is the process of sampling a sound so that it can become a series of bits (0/1) rather than the normal series of bytes, which is a massive memory saving.

However, this encoding introduces noise and loses quality. The DPCM algorithm was originally meant for compressing human speech but works well on percussion and drums that rely on punch and crunch rather than high fidelity.

That said, numerous games including the Konami game *Super C* (1990) make good use of tuned DPCM samples to play instrumental parts.

The channel can also play uncompressed PCM samples. This means better quality, but the consequent CPU load means it's best suited to title screens, such as Rob Hubbard's *Ski or Die* which uses crunchy sampled distortion guitars.

Wild Duck Hunt Master

Games using the sample voice include *Wild Gunman*, *Duck Hunt* (dog woof and duck quack), *Kung-Fu Master* (known as *Kung-Fu* on the NES) and 1988's *Gradius II* (Japan only, voice samples).

Approved developers only

Except for cartridges that allowed the consumer to compose limited tunes, the wonders of APU were hidden from the public until emulation and tools were available for hobbyists and demo sceners to experiment for themselves, Nintendo lawyers permitting.

Like the C64 (.sid) and the AY-3-891x (.ay), NES music now has its own file format (.nsf). NES tunes can be distributed standalone and played on other systems, and even created on other systems in tools such as *FamiTracker*. Of course, these tunes can also be played on real hardware.

Arcades and clones

In the arcades, most deployment was through the PlayChoice-10, which brought regular NES games to the arcade (e.g. *Super Mario Bros*).

There was also the Nintendo VS. System which featured games such as *Vs. Super Xevious*.

The success of the NES and the greed of humans made clones inevitable. A particularly successful one was the Micro Genius IQ-501/502. This is more of a Famicom clone than an NES one, needing an adaptor to play NES carts, but it is compatible with the extra audio from enhanced cartridges.

It's slower though, thanks to the difference between European and Japanese TV standards (50 vs 60 Hz).

The IQ-501 was sold as "Dendy Classic" in the Soviet Union, becoming legendary in the process. Less legendary are other '90s clones such as the "Golden China TV Game" which was, surprisingly, quite popular in South Africa.

There were also a number of 1990s "Keyboard Famiclones" such as the Subor SB-486. These seem an uncomfortable mix between console and computer, though English language learning software was popular and there was at least one model aimed at children ("Cyber Computer").

Irony

Strangely, the chip was deployed in the first Atari Flashback device in 2004. This necessitated a series of very quick ports from the 2600/7800 platform to this NES-on-a-chip, with varying degrees of success.

This was ironic because Atari had the opportunity to market the NES in the USA but failed to seal a deal. So, this led to Atari's history being preserved through a Nintendo device. Later Flashback releases were truer to the original hardware and games.

Musical Heroes

The occasional visitor to the Nintendo universe would have heard of Kōji Kondō, composer for the *Super Mario Bros* and *The Legend of Zelda* franchises.

However, NES chipmusic fans, while respecting his classic works, reserve much of their enthusiasm for other composers such as Manami Matsumae (*Mega Man*), Yasuaki Fujita (*Mega Man 3*) and Harumi Fujita (also *Mega Man 3*). If you're thinking that the *Mega Man* franchise is slightly over-represented in that list, you probably haven't seen a top 10 list of NES soundtracks lately! Kenichi Matsubara's *Castlevania II* music is also highly rated.

Western composers to make a mark in NES-land include Tim Follin and Jeroen Tel.

NES/Famicom Expansions

They just kept expanding Famicom in interesting ways… but sonically left the NES in the cold.

Famicom Disc System

This was a floppy-disc add on, released in Japan, February 21st, 1986. Normally that would be uninteresting sonically, but it also delivered the Nintendo FDS wave generator - an ASIC chip named the 2C33 for single voice, single-cycle wavetable sound.

Only one more voice? Surely no one would bother using it? But they did. The list of games released with enhanced soundtracks for the FDS is lengthy, including big names such as *The Legend of Zelda*, *Doki Doki Panic*, *Castlevania* and *Metroid*.

In some cases, the extra voice made the music richer, and in others the soundtrack had been reworked. *Metroid*, for instance, replaced toggled one-voice chords with solid three-voice ones. This required a thorough revamp of the entire soundtrack and some strategic musical decisions.

Famicom Controller

For the sake of completeness, we must mention that the Japanese Famicom's second controller features a small condenser microphone. The audio input from that can be mixed with the main audio and can also control features within games.

Its original function was for karaoke, but its most famous use turned out to be defeating a particular enemy in *The Legend of Zelda*. It also featured in *Raid on Bungeling Bay* and *Star Soldier* and could even be PEEK'd in *Family Basic*.

The microphone was removed from the NES version, but recently made a glorious return on the re-released Famicom controller for the Nintendo Switch.

Cartridge Expansion chips

The NES/Famicom was a hugely popular platform with a massive user-base. This meant that companies felt they could afford to support more obscure parts of the Famicom tech.

In this case, the audio line from the cartridge port allowed cartridges to contain their own additional sound chips to augment the APU.

However, the Western NES moved that audio connection from the main cartridge port to the expansion port at the bottom of the console which, it turns out, was hardly ever used (except by a hacker who used it in 2015 to send tweets). It's almost as if the expansion port is the "real" Famicom cartridge port, and it's the NES cartridge port that's the add-on!

This loss meant that NES versions of audio-enhanced Famicom games needed to be adjusted in order to avoid losing voices. However, surprisingly few of the enhanced Famicom games were ported.

Konami VRC6
Used in Akumajou Densetsu (Castlevania III: Dracula's Curse), Madara and Esper Dream 2.

Apart from providing a screen mapper to greatly improve graphics, the VRC6 provides three extra channels for sound: two pulse waves, and one sawtooth.

All channels operate similarly to the native channels in the NES APU. This only works on the Famicom of course.

Castlevania III's NES release shipped with a Nintendo MMC5 chip rather than the original Konami VRC6. This probably wasn't a sound-related change because neither chip can deliver enhanced audio on the stock NES. It's more likely to do with PAL compatibility, the official MMC5 offering greater compatibility in that area.

Nintendo MMC5
Used in Castlevania III: Dracula's Curse (North America/EU), Just Breed, Metal Slader Glory, Laser Invasion, Uchuu Keibitai SDF, Nobunaga's Ambition II, Nobunaga no Yabou - Sengoku Gunyuu Den, Bandit Kings of Ancient China, Romance of the Three Kingdoms II, Uncharted Waters, Genghis Khan II: Clan of the Gray Wolf, Gemfire, L'Empereur, Ishin no Arashi, Shin 4 Nin Uchi Mahjong - Yakuman Tengoku.

This powerful (and expensive!) screen mapper chip provides extra sound output like the VRC6, but with the sawtooth wave replaced by a PCM channel. As would be expected for a chip required to blend in with another chip, the volume levels are matched, though the sample channel can be boosted to become twice as loud.

Most games with this chip use it for enhanced graphics, not sound. However, *Just Breed* and *Shin 4-Jin Uchi Mahjong - Yakuman Tengoku* used the chip musically to great effect. A few games use the audio for SFX: *Ishin no Arashi*, *Metal Slader Glory*, *Shin 4-Nin Uchi Mahjong* and *Uchuu Keibitai SDF*.

Konami VRC7
Used in Lagrange Point (Konami) and Tiny Toon Adventures (sound expansion not used).

Aside from the graphical enhancements, the VRC7 is notable for shipping with an even more cut-down version of an already cut-down chip: the YM2413 OPLL. This has 6 channels of 2-operator

FM and implements a subset of the YM2413, alongside a bespoke (fixed) set of instruments.

Konami skipped around the potential problem of porting *Lagrange Point* to the NES (and losing its music) by not actually porting the game. At least all NES users can now play an emulation though!

Released late in the Famicom's life, this soundtrack is widely regarded as a classic, and it's a shame that the VRC7 didn't see other artistic use.

Sunsoft 5B
Used in Gimmick! Seen in other cartridges such as Gremlins 2.

The Sunsoft 5B is part of a chip family called FME-7, which also includes a 5A variant. The 5B's claim to fame is enhanced audio.

The idea behind it is lovely: a combination of the NES chip and the Yamaha YM2149F, as made famous by the Atari ST! 8-voice-SSG-tastic!

The 5B chip contains an integrated YM2149F core rather than featuring the chip itself, allowing it to fit into the space required on the cartridge PCB. The YM's natural volume is louder than the NES's APU, so that needs to be balanced by the musician.

The only game that uses this extra audio is *Gimmick!*, which is generally recognised as one of the best music soundtracks on the Famicom.

Renamed *Mr. Gimmick*, this game was planned for worldwide NES release, but only saw the light of day in Scandinavia. It has reworked APU sound, and the cartridge contains one of the Sunsoft chips without the audio enhancements. This European cartridge is now one of the rarest NES cartridges in the world.

Some other Sunsoft cartridges that use FME-7 or *Gremlins 2* cartridges have been found with 5B chips in them, even though they didn't need the extra audio. These cartridges tend to be coveted because of the mere chance they may contain a rare 5B chip.

Namco 163
Used in Final Lap (4 channel), Rolling Thunder (4 channel), King of Kings (8 channel).

Ironically, this 1988 chip wasn't used on any Namco Famicom games, but only on third-party games that Namco published without Nintendo's authorisation.

It's a powerful 8-voice chip using 4-bit variable length wavetables. Only one composer (Tsukasa Masuko) made full use of all eight channels, most composers being content with using four of them.

Unlike most sound chips, there's a trade-off between the quality of the output and the number of voices in use.

The chip has 128 bytes of RAM. This is shared by the channel registers and the samples. With 64 of those bytes devoted to samples, that's eight 16-step waveforms.

If you use fewer channels, your samples can be larger and more detailed because they each have more RAM.

Also, the mixing algorithm cycles through the voices very quickly to give the illusion they're all played at once. Fewer active voices means more resources for each one, resulting in better audio quality.

Even with only four voices in use, the combination of this chip and the APU gives rich 9-voice chipmusic.

Luckily for gamers, the few 163-based games that did gain a NES release such as *Rolling Thunder* had their soundtracks rewritten and simplified for the NES chip rather than just losing channels.

Common chipmusic tricks such as note-flipping/multiplexing for chords were used to fill musical gaps left by the lack of voices.

Jaleco NEC μPD7755C/μPD7756
Used in The Bases Loaded Series (Moero!! Pro Yakyuu onwards), Pro Tennis/Racket Attack, Moero!! Pro Yakyuu, Moero!! Pro Yakyuu '88: Kettei Ban, Moe Pro! '90: Kandou-hen, Moe Pro!: Saikyou-hen, Moero!! Pro Tennis.

Jaleco's Japanese baseball games for the Famicom found an easy home in the baseball-crazy US, but the NEC ADPCM speech chip in the original cartridges didn't travel with it.

When the NEC chip and the special ROM containing the samples disappeared from the American version, they were added to the program ROM to play through the NES sample channel.

Mitsubishi M50805 (Bandai)
Used in Family Trainer 3: Aerobics Studio.

The third instalment of *Family Trainer* somehow required an 8-voice wavetable chip allowing 120-byte ADPCM 4-bit samples, which is more than enough to shout at you for slacking! This was ported to the NES as *Aerobics Studio*, and presumably got less vocal in the process.

Japanese vs European Chipmusic

Two types of chipmusic developed very differently. Why, and how?

Mike Clarke, Psygnosis alumnus, explained to us his thoughts on the root cause of the differences in chipmusic culture:

> *"Japanese chip music as far as I know was, like in the USA, not very technical for the most part. They were mostly classically trained musicians rather than programmers. There was a lot of usage of MML and later MIDI for music data. The programmers would write a player, the musicians would write the music data.*
>
> *This led to fairly rudimentary music sound-wise, mostly consisting of straight waveforms. I remember seeing one of the documentaries where the musicians could choose from a bunch of preset waveforms for the sample channel on the NES. These had just been created by the programmers of the play routine.*
>
> *If you think about it, it couldn't really have been any other way. There wasn't much of a computer game industry because almost everything was arcades and consoles, so to be able to write game music you had to have a devkit, which cost thousands, and a development license, which was nigh on impossible to get if you weren't an actual company.*

Therefore, to create music, you had to first get a job at a game company. Unlike in Europe, you couldn't write your own play routine at home. The barrier to entry was way too high. The few actual programmer/musicians like Yuzo Koshiro really stood out.

The Americans have this weird fetish about Japanese game music on the home consoles when in reality it wasn't really very good. There's not really anything particularly impressive until the Mega Drive and even then, there are only a few standouts like Streets of Rage and Sonic."

Cultural differences

The method by which someone becomes a game composer has a profound impact on the types of people hired, and the music they produce.

In Europe, teenager programmers with a musical inclination impressed companies and were hired, regardless of musical training. In Japan, not so much.

Points of Difference

Other points of difference surface in reviews of Western tunes that somehow wander into the lower reaches of some NES popularity charts, such as "Top 100 NES Tracks as voted by The Shizz".

Their top reviewer is "Jace", an expert in the arcane ways of differential chipmusic!

Here's Jace reviewing Neil Baldwin's NES port of Barry Leitch's *HeroQuest* theme:

> "We could (and someone probably should) write an entire essay on the innovations and accomplishments of Neil Baldwin.
>
> The HeroQuest soundtrack is one of his better-known works, defined by appropriately 'medieval' sounding scales and chord progressions juxtaposed with driving, modern rhythms, and masterful volume manipulation to achieve gorgeous delay and echo effects.
>
> Mr. Baldwin is also a master of vibrato; largely through this tool, his tracks have a warmth and humanity about them that is rare on the NES."

His review of Jeroen Tel's *Overlord* takes in the advantages and pitfalls of warbling arpeggios:

> "The melody is nothing too memorable, but the bubbling arpeggiated chords flowing underneath are perfectly programmed into gorgeous washes, creating a distinct and non-disruptive atmosphere.
>
> This typically European approach to square wave manipulation is not generally favored at The Shizz, but it can be quite impactful when tastefully used as a background layer as in this creative Overlord track."

Alberto Gonzalez's banging "The Flight on a Stork" from *The Smurfs*, couldn't be further away from the usual NES fare.

According to Jace:

> "Aside from this song being awesome enough to make me flip my laptop off my lap and punch myself repeatedly in the face on the quarter notes the first time I heard it, the best thing about it is probably the reaction you get when you're listening to chiptunes and VGM around the one or two friends you have who can tolerate them and you tell them that this epic dance floor crasher is from a Smurfs game.
>
> The Amiga and C64 catalogs are full of these kinds of unbalanced corny game/slammin' tunes dichotomies, but it's a bit more rare on the NES and extreme in this case."

Crossover

> "The crossover between the two styles was Tim Follin. They love the Jazzy styles in Japan. I believe he later influenced them into that style via several of the Software Creations games that he did for Nintendo.
>
> Japanese music on Japanese consoles has mostly been really good - Mario, Castlevania, Mario Kart, Star Fox, etc."

Allister Brimble

Follinnovation

If Tim Follin is the crossover point, then detailed musical reviews are a great way to pinpoint the differences and similarities between Tim's work, and other legendary work such as *Mega Man 3* or *Castlevania*.

SID-like tricks were often an acquired taste with audiences outside Europe, but they loved Tim's.

Jace's *Silver Surfer* review sums it up:

> *"Many NES composers, Tim Follin foremost among them, have an uncanny knack for emulating actual instruments with the NES, and most commonly it sounds like people were shooting for the electric guitar. Well, nowhere else does a 2A03 square wave sound more like an electric guitar than in this song, particularly with that single piercing high note at the end of the brief intro.*
>
> *But Mr. Follin wasn't content with merely emulating the guitar. I believe that every single programming trick possible is pulled out in this song, giving it an explosive pizzazz quite unlike anything done for any other platform in the 8-bit era, and generally not sounding quite like anything in purely synthesized music for that matter.*
>
> *The greatest Follinnovation of all is the use of the triangle wave to handle two things at once: drum kit and bass. Primarily because of the drum sound, I still can't fully wrap my head around the fact that this song contains no sampled content whatsoever, nor any expansion chips. Furthermore, there's a sense of loudness to this song that should be impossible on paper.*

The 2A03's triangle wave had a fixed volume, which should mean that pushing the squares too hard would drown it out, but EVERYTHING in this track is blowing up in your face, bursting with an up-frontness that is beyond my comprehension.

Deliberate application of certain frequencies likely has something to do with this perceived loudness, as well as the specifics of the interplay between square wave velocities and the fixed triangle.

The greatest criticism of Follin is that he tends to eschew melody and structure in favor of technical details, kitchen-sink flash, and blinding shred. A lot of times it sort of sounds like he's just having a self-indulgent laugh. That criticism doesn't hold with this particular song, though. Sure, it has the appearance of one billion things crashing into each other in a spray of neon glitter all shooting out of Galactus' mighty horns, but somehow, against all odds, this one manages to get stuck in your head.

I'm not necessarily saying that 'good music' needs to be catchy, or that I'm talking about catchy in a Super Mario Bros. way, but this particular creation is all the more amazing for its melodic choices in tandem with insane chunks of flair and the boldness of its chord changes."

(1983) CEM3394 Synthesizer Voice

Unique analog-synth-on-a-chip that was better than the music it had to play.

As seen in: Bally/Sente Cabinets

1-voice SSG. Mono.

We know that one of the original hopes of Bob Yannes, SID chip designer, was to create a chip that could also make the journey into professional studios.

The Curtis CEM3394 is an example of a chip that came the other way, having been designed for synthesizer use by established synth technology company Curtis Electromusic Specialties.

While there were numerous games using the Bally/Sente platform, there was only one type of cabinet and the game ROMs were purchased on plug-in boards, an idea ahead of its time. The audio PCB was built into the base hardware and not into the game boards. This meant that all games using this hardware had access to six CEM3394s and a noise generator whether they needed it or not.

Wave Control

The last chip we looked at that had an analog component was *Space Invaders*' SN76477, which was not designed with microprocessor control in mind.

Equally, the CEM3394 needs an extra layer of silicon to allow a digital microprocessor (in the Bally/Sente cabinet, a Z80A) to take advantage of the features in the CEM3394.

This layer is an analog to digital switch, one for each CEM chip.

The chip used is a 4051B Analog Multiplexer/Demultiplexer (the chip in the image below is labelled MC14051B but is still a 4051B!).

Its job is to take digital signals as input and output the voltages that the CEM3394 requires to function.

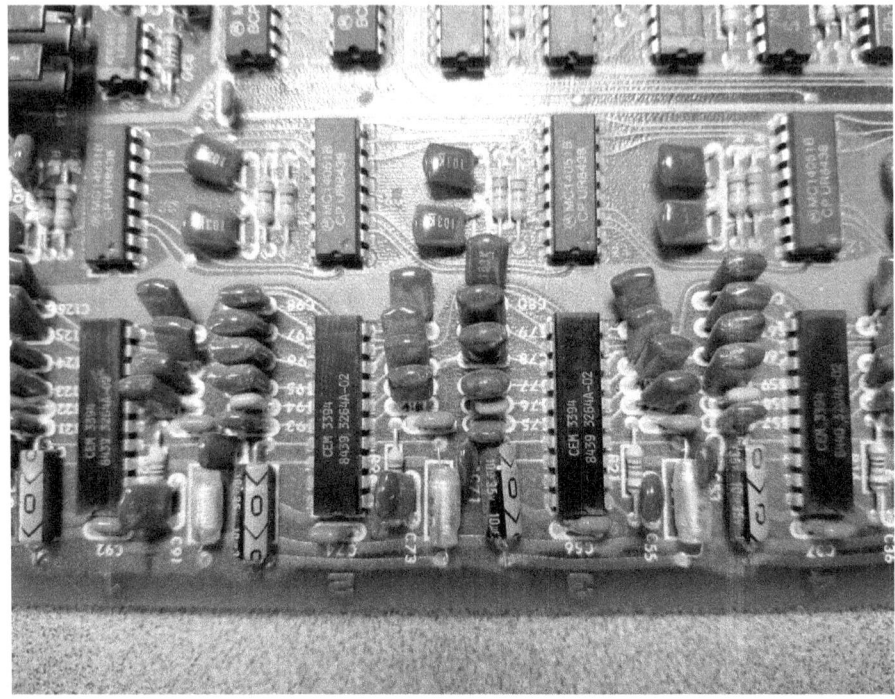

Photo courtesy of cpdist.com

Waveform Selection

Pin 6 Voltage	Triangle	Sawtooth	Pulse
< -0.45	Off	Off	On
-0.25 to +1.0	On	Off	On
+1.4 to +2.3	On	On	On
> +3.1	Off	On	On

Waveform selection, for example, is achieved by sending specific voltages to pin 6 on the CEM3394 via the C14051B.

The chip's architecture for mixing the different oscillators (pulse, triangle, sawtooth) ensures rich sounds can be produced even though the chip itself can only play one note at a time.

The pulse wave is always "on". To hear the triangle and/or sawtooth waves in isolation the datasheet suggests sending a slightly negative voltage to pin 7 (the pulse-width control) to set the pulse-width to zero and make it inaudible.

The chip also tweaks the overall loudness of the waveforms to ensure they sound balanced.

It makes the triangle wave ~27% larger than the sawtooth wave, and the sawtooth wave ~27% larger than the pulse wave.

Analog Love

This is a very lovely chip indeed, and emulation in MAME hardly does justice to its analog smoothness.

So why is analog technology a good thing? It has its drawbacks: it's less predictable and tends to drift out of tune when warmed up. In fact, the CEM datasheet boasts a lot about heat stability of this chip!

However, analog oscillators/sounds and especially analog filters can be smooth and full, partly due to the organic variation in voltages, and because physical materials can be made to behave in non-linear ways that it took decades for digital technology to properly replicate - for instance, with virtual analog synth technology (though many musicians still insist they can hear the difference).

Great though it is, one chip plays only one note. That's why six chips were deployed in the Bally/Sente system, giving six-note polyphony.

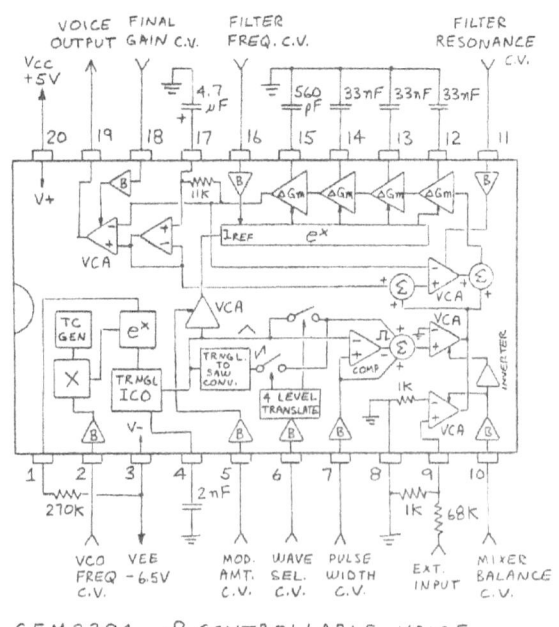

CEM3394 µP CONTROLLABLE VOICE BLOCK AND EXTERNAL CONNECTIONS DIAGRAM

Sounds like SID?

The chip shares a lot of superficial and audio similarities with the SID, though that chip is a digital chip with an analog filter, and the CEM3394 is fully analog.

The key similarities to the SID are its support for micro-control of the duty cycle of the square wave (making "fizzy" pulse-width modulation possible), and support for both triangle and sawtooth waveforms.

While both chips allow selecting more than one waveform at a time, the CEM mixes the oscillators together to create a combined sound, but the SID combines the waveforms mathematically in a much more complex way.

Both chips also have a powerful filter (mostly used for making sounds smoother), and a resonance setting making "weeeee" sounds possible.

However, on a SID there's one filter shared between the three voices, whereas each CEM3394 can have its own filter settings.

One key difference from the SID is that the individual CEMs on a board cannot modulate each other. However, SID voices can interact to great creative effect using features such as "ring modulation" and "hard sync".

One giant disadvantage for a video-game chip is the lack of a noise waveform, but an integrated NEC MM5837 Digital Noise Stream fills that gap.

Like the SID, the CEM3394 has an external input for sound from other sources. This is useful for daisy chaining the output from the multiple CEM3394s in the system, and for adding in the additional noise from the MM5837.

Unlike the SID, the chip has no hardware envelopes to change the shape of the sound. This could be accomplished with the add-on chip CEM3371 ("I'll have six please!") or using the CPU to modulate the volume of the sound over time. Bally/Sente's cabinet does not have CEM3371s, so this job was left to the Z80A.

Great tech, shame about the music.

So how was this lovely miracle of analog engineering used?

The technology was often better than the compositional skill of the programmers using it - so while the music always had a pleasantly SID-like sound, it was sometimes unchallenging and over-jolly, though later games such as Rescue Raider made a better job of it.

What synths contained a CEM3394?

- *Sequential Circuits MultiTrack*
- *Sequential Circuits Max*
- *Simmons SDS 800 (1984)*
- *Sequential Circuits Six-Trak (1984)*
- *Sequential Circuits Split 8 (1985)*
- *Simmons SDS 9 (1985)*
- *Akai AX 73 (1986)*
- *Akai AX 60 (1986)*
- *Akai VX90 (1986)*

Find one today!
Though you might need to re-mortgage your house!

(1984) Atari AMY

Amazing additive sound-chip for the Atari ST murdered by corporate shenanigans.

As almost seen in: Atari ST/65XEM.

8 voices based on additive synthesis (sine waves and fourier transform) with additional noise generators. Stereo output.

PAULA killer?

> *"AMY, stands for Additive Musical sYnthesis"*
> **Leonard Tramiel**

AMY is one of the most intriguing synth chips ever developed. It might even have tilted the Amiga/ST battle in a different direction had it been implemented in the 520ST as planned... or, maybe not, given how idiosyncratic the chip was and how the Amiga would still have outclassed the ST in custom graphics hardware.

By 1983, the trend in synthesized audio was away from the "dated" analog synthesis that had been around for decades, and towards a quest for more realistic sounds.

Rapidly becoming more affordable, cutting-edge synth technology such as the bright and punchy FM synthesis had entered the arcades.

Newer and more powerful chips for sample playback were also coming to market.

A new realism

The synthesis in most sound chips until 1984 was based on "subtractive" synthesis. You start with a very "rich" waveform (like a square wave), and then you shape, filter, and modulate it to approximate what you want. It's like sculpting with sound.

However, another approach is "additive synthesis" - the idea that all complex waveforms are made of a very complex interaction between many different sine waves.

The method used to break down a complex waveform to its component waves is called "Fourier Analysis". Additive synthesis means you can represent a complex wave in high precision with a small amount of data to describe it, which is the perfect recipe for realism.

AMY jumped on this with gusto, and threw in an extra noise generator as well, because everyone needs a good shift register.

Contract killing

This chip never made it to consumers because of complicated legal problems between Atari and a third-party they hired to develop the chip from their initial design - which started when the company did indeed develop the design, then refused to licence it back to Atari.

In retrospect, this pure synthesis chip would have been glorious but at odds with the way the market and the industry were heading, which was towards "realistic" synthesized instruments from FM chips, or increasingly affordable reproduction of sampled sound. Composers would probably have ported over a tracker for it and turned it into a sample player, with only a few using the chip to its potential. This was the same fate that befell the ES5503 chip in the Apple IIGS.

So, the Atari ST line launched with the YM2149F, a licenced AY-3-891x variant. It's good to know they tried to give it a better chip first though.

The strangest part of this story is that Atari subsequently designed the chip into the existing 65XE Atari 8-bit computer to produce the 65XEM.

It's difficult to see any logic except desperation behind this decision, which would have created a complete mismatch of two different levels of technology. This model was also subsequently cancelled.

(1984) Yamaha YM2149F

Yamaha's rebadging and tweaking of the AY-3-8910 extended its lifespan by years.

As seen in: Atari ST, MSX2 and a squillion Mahjong machines.

3 square-wave voices, with noise and ADSR shared between the three voices. A full subsonic to supersonic note range, and a low-pass filter. Noise is from a 17-bit linear feedback shift register. Individual voice outputs have their own pinouts and can be mixed to mono or stereo depending on implementation.

The same but different

Wait a minute, isn't this the same as the 1978 AY-3-8910?

Well, it's pin-compatible (for chips, the best kind of compatible). There are a couple of minor differences: pin 26 can halve the master clock if pulled low, and the generated signal has a 2V DC component, as opposed to 0.2V DC with the AY.

But does it *sound* different?

Some do report it as sounding slightly different to the AY. AY/YM expert Sergei Bulba had this to say in a post at the English Amiga Board:

> "For [the] listener AY is a little bit louder than YM, because YM has some steep DAC output graph (YM is closer to logarithmic, than AY). Therefore some music melodies on AY can be called 'legato' and on YM (are) 'staccato'."

One reason for this might be the envelope generator having twice the number of steps and operating twice as fast.

Ben Daglish was more flippant about the differences between chips, joking in an interview with long-time ST user Richard Karsmakers:

> *"I prefer AY-3-8912 because the sub-modulated square-generated harmonic frequency distribution was more to my liking!"*

The YM2149F replaced the AY-3-8910 chip in the MSX2 specification and appeared in the Tandy Color computer.

Any port in a STorm

The most high-profile deployment of this chip in the home was in the Atari ST computer family, where it was a last-minute substitute for the amazing AMY chip that met an unfortunate legal demise.

Having a sound-chip functionally identical to the ZX Spectrum 128 and the Amstrad CPC makes it *very* tempting for a composer to convert a Z80A driver to 68000 assembly and port the same chunk of music data from one platform to another, with near-identical results.

Quite often an identical tune sounds more stressed or lively on the ST than the Spectrum 128 because it's slightly higher pitched. This is because the note table frequencies may not have been updated in the source code to account for the fact that the AY on the Spectrum runs at 1.77 MHz and the YM on the Atari ST runs at 2 MHz.

STars

Composers David Whittaker and Jochen Hippel were by far the biggest players in ST game releases, David also being a prolific game composer on the C64, Spectrum, Amstrad, and Amiga.

He also produced a variety of types of tune, so *Beyond the Ice Palace*, a direct port of his AY Spectrum 128 version, sat alongside big budget sample-filled epics such as *Shadow of the Beast*.

Jochen's work often sounded much more SID-like and bouncy than AY music, though he could AY blip with the best of them. But he also developed the technology to play 4-channel Amiga MODs back through only one voice on the ST. This technology was later backported by Chris Huelsbeck to the Amiga, resulting in 7-channel MODs!

Jochen was also responsible for porting most of Rob Hubbard's music to the ST in demo form, using his demo handle "Mad Max of The Exceptions (TEX)".

It's not just the chip, you know...

Jochen's work illustrates how game music isn't just about the sound chip. Composers had more opportunity (CPU and RAM) on more powerful computers to be more adventurous with their soundtracks if they chose. The ST was a particularly powerful partner if the composer wanted to go beyond the limitations of the AY core.

Jochen often did his utmost to make the ST sound like an Amiga or a C64, and David Whittaker had sonic triumphs on the platform in some big-budget games. David was the king of the AY sound, but he also knew how to deploy impressive sample-driven OSTs.

Shadow of the Beast (the work he's most known for worldwide) featured much the same instruments and vibe as the Amiga version, which majored in pan flutes and atmosphere. He also worked with Bitmap Brothers on *Xenon 2: Megablast*, deconstructing Bomb the Bass's track "Megablast (Hip Hop on Precinct 13)" to use as the game's licenced soundtrack.

In his later career he worked for software house Traveller's Tales, doing similar work on the best-selling LEGO® games.

David Whittaker the trickSTer

Even in his AY-inspired work, David had some clever tricks up his sleeve.

The most audible one is the "AY Echo". This is giving a musical part a distinct delay/echo sound despite the AY/YM chips not having any built-in effects processing.

While it's easy to produce an echo in two voices (by playing the same thing on another channel, but slightly quieter and later), David, by contrast, only needs one channel. He tricks the ear by repeating the start of a note after a very short time, but at a quieter volume.

The crucial trick is that David's code doesn't restart the note's vibrato when it plays the repeated note. This continuity convinces the ear that there was only one note played.

For fast arpeggios, David was even cleverer, as seen in *Glider Rider*. The speedy arpeggios play three notes, then replay two of the notes at a quieter volume level. This results in a shimmering delay.

Similar tricks were later used by other composers such as the legendary Tim Follin. The principles aren't specific to this chip, so they can also be used on other sonic platforms.

STandalone

In its standalone form, this chip had a large deployment in coin-ops, though it was more likely to appear in games such as *Hi Lo Casino* than in the big releases of the day. In MAME's list of games deployed with the chip, the most famous is *Buggy Boy*, the composer of which stays frustratingly and stubbornly anonymous.

In coin-ops, the YM2149F was quite often deployed along with FM chips, potentially being useful as a noise generator, a source of extra instruments or even a source of clock signals.

The SSG core of the chip did achieve wider exposure as a module in other Yamaha chips, usually overshadowed by the FM and PCM modules. It sounded great when it was used properly, and a lot of PC'98 music makes good use of the oscillators.

I have to give a shout-out to the company Kaneco at this point: in the early '90s while arcades were in their FM phase, Kaneco persisted with two YM2149s and paired them with an MSM6295 (for sampled drums) to produce some heart-warming work on games such as *Magical Crystals*, *The Berlin Wall*, and *Explosive Breaker*. A real commitment to quality chipmusic!

A demo scene worth chronicling

The ST demo scene, as chronicled exhaustively in three volumes by author Marco Breddin ("The Atari ST and the Creative People") was an opportunity for creatives apart from Jochen Hippel to push the technology to its limits.

Marco's suggestions for notable demos are in the gallery that follows, and he also recommends *The B.I.G. Demo* by TEX because of its huge library of Rob Hubbard conversions by Hippel himself.

AmazeST

Other amazing demos include: *Mindbomb Demo, Life's A Bitch, Overscan Demos, European Demos, Decade Demos, Anomaly Megademo, Eat my Bollocks* and *Flip-O-Demo*.

For more modern demos, you should also check out *Suretrip 49% Version* by Checkpoint, *Sea of Colour* by Dead Hackers Society (for the Atari STE), and *STNICCC 2000* and *We Were @*, both by Oxygene (STE).

Also, there's a special mention for *Decade Demos* with Count Zero having a "unique chipmusic style". The musician Tao is also recommended for his chip style, best experienced in the music demo *Steps*.

His modern tunes sound almost like samples even on a stock YM2149F, and he also made a 20-minute tune covering the development of ST chip sound over the years: *Ultimate Muzak Demo 8730*. Historical!

> *"It might also be worth mentioning ST sound programmer Holger Gehrmann who was the first to get double-tone mode on the ST, quite early in the machine's history.*
>
> *All his tunes were brought together in the demo Synth Sample V.*
>
> *The first game with this at the time ground-breaking music was Extensor, which was also special because it worked on monochrome as well as colour monitors (there weren't many of these because most games only worked on colour monitors)."*
>
> **Richard Karsmakers, ST MegaFan**

(1984) Yamaha YM2151 (OPM)

This FM chip and its descendants defined video game music for nearly a decade.

As seen in: Marble Madness, OutRun, CX5M (MSX1), Sharp X68000, Capcom CP System.

4 operators per channel over 8 FM channels. Composite Sinusoidal Modelling. Stereo output.

A new engine

Just as the SID was the first chip to introduce a fully featured subtractive synthesis engine into an off-the-shelf digital sound chip, the YM2151 was the first single-chip implementation of Yamaha's FM technology.

First introduced in 1980, this is the same technology powering the legendary DX synthesizers that provided clanky basses and tinkly electric pianos for a plethora of 1980s hits.

The chip has a full 8 voices, each with four FM operators available. This makes it more powerful than the popular YM2203 OPN chip which has three FM voices and a YM2149F SSG core.

However, it's not as well-specified as Yamaha's legendary DX7 standalone synthesizer, which allowed six operators per voice. This difference in specifications meant that some complex DX7 patches are impossible to reproduce in their original form on the YM2151.

As you can tell from the above jargon about operators, there's no getting round the fact that FM synthesis is difficult to understand. Most musicians learn by experimentation or just use slightly modified presets.

However, *this* chip has no presets. Its first deployment in *Marble Madness* saw Brad Fuller and Hal Canon spend months without English documentation experimenting with registers until they had something they liked. It was also the first stereo coin-op.

Atari's license with Yamaha meant they had the chip to themselves for a year, time which they used to create some unique soundtracks. However, from 1985 the deluge started as the YM2151 found its way into a plethora of arcade cabinet releases worldwide.

Magical Sound Clang

However much you experiment, the sound of FM is distinctive, typified by the middle-of-the-road sounds of *OutRun*'s "Splash Wave" or "Magical Sound Shower" or even, in a more basic two-operator form, the Adlib music found in classic PC DOS games.

So, FM usually means clanky sounds, whiny sounds, and tuned percussion, with bells being a specialty. Earl Vickers did manage to make the YM2151 speak in *Xybots*, using an obscure feature ("composite sinusoidal modelling") to its fullest.

FM would stick around for a while, finally getting to a mainstream games console with the launch of the Sega Genesis/Mega Drive in 1988 but proliferating in PCs and musical add-ons before that.

Also seen with...

FM was new and exciting, but it wasn't good at everything, and had limitations that needed addressing. Sample playback and speech synthesis chips soon became an essential addition to the YM2151, satisfying the need for drums, bass, and orchestra hits, or just extra voices.

For example, Atari's classic *Gauntlet* sported an additional POKEY chip (for legacy sounds and the POKEY's I/O functions) and a TI TMS5200 speech chip.

In iconic Sega System 16 games such as *OutRun*, it was deployed alongside a SegaPCM chip for the speech and sound effects, and bouncy drums.

The FM engine of the YM2151 is not known for its ability to create convincing rhythms. Therefore, this job was often delegated to PCM or ADPCM chips, making the music sharper and punchier and reducing the voice load on the YM2151 so it could play to its melodic strengths.

Later, trends in pop music resulted in every game seemingly feeling compelled to use orchestral hits, which is a sound that requires a sample playback chip to be convincing.

This configuration of chunky 8-voice FM music and crunchy samples/speech defined the sound of the arcades in the mid-to-late 1980s.

Konami also got in on the FM action, using this chip in notable arcade games such as *Gradius II*, *Super Contra* and *Teenage Mutant Ninja Turtles*, where it was combined with a μPD7759C sample playback chip and a 007232 Konami Custom Chip, with the load taken off the main CPU by a venerable Z80A, the most common audio CPU in arcade deployments.

Capcom also used this chip for its CP System launched in 1988, which featured iconic games such as *Forgotten Worlds*, *Strider* and *Pang*. It was paired with another sample chip, the MSM6295 (also called the OKI6295) which could handle speech synthesis and any human-originated blood-curdling death sounds required.

YM2151 - Home Life

The chip did make it into the home. As the first of its kind, it was brand-new flagship technology, so it wasn't cheap to third parties, but Yamaha themselves used it in their CX5M MSX music computer, with its custom full-size keyboard and promise of full synthesizer sound.

The other deployments were from Sharp in their X1 Turbo line and its successor, the X68000. This system launched to consumers with *Gradius* and was used as the development rig for Capcom's CP System. Released in 1987 it was expensive (over $4,000 in today's money), but it was a genuine arcade experience.

For that price, it would have to be!

(1984) Yamaha YM2203 (OPN)

Hybrid FM/PSG chip best known for its appearance in Space Harrier.

As seen in: PC-6001 Mk II SR/PC-6002, PC-8801mkII, ALPHA-68K96II, and many coin-ops.

SSG core is from YM2149F. FM has 3 channels, 4-operators per channel. In use, voices 1-3 are SSG, and 4-6 are FM.

The migration to FM begins

With this chip, Yamaha began the process of migration from SSG to FM by grafting 3 new voices of FM onto the older YM2149F core.

One quirk to mention here is that the AY sounds come out of the chip as analog (no DAC needed), and the FM sounds come out as digital (needing their own DAC), which means having to mix the sounds off-chip.

There's something wrong on the Internet!

The chip's first use was in the NEC PC-6001 Mk II SR, and it became a stalwart chip in the entire NEC PC line from 1984 to 1991.

Finding information about this chip highlights one of the problems of researching a book decades after the fact.

In this case, it was challenging to confirm the presence of a YM2203 in the PC-6001 machine. Several internet sources are simply incorrect and insist that this PC just has a YM2149F chip. It was also difficult to find PCB photos with enough detail to confirm the presence of the chip at all.

Star witness

No matter how niche the computer, there is always someone for whom it was their first love, and so first-hand witness evidence was achieved.

In this case many years after initial release, a PC-6002 was released in Iraq and ended up in the hands of 11-year-old Salwan Hilali, who had the choice between this and a used Amstrad CPC464.

None of his friends were familiar with the CPC, so his first computer was this exotic Japanese model.

Like many owners of home micros, their computer proved life-transforming. In this case leading to a lifetime in video game development.

> "EIC (Electronics Industry Company) produced two lines of personal computers both licensed from overseas manufacturers. The first was the Al-Warka PC based on NEC's popular Japanese PC-6001 series and targeting enthusiasts and students.
>
> N66 SR BASIC provided access to an FM chip: Yamaha YM2203 in addition to the PSG core. They were both usable. So, you can actually play six tracks at the same time: 3 FM and 3 PSG.
>
> N66 SR BASIC came with 12 FM preset instruments, and it also allowed programming your own instrument if you knew the right combination of magical POKEs to the right addresses."
>
> **Salwan Hilali, zenithsal.com**

The story of home micros and their often-unofficial clones outside Western Europe, North America and Japan is worth a book, but has been covered in numerous articles, especially the Eastern European computer models.

Fun fact: one videogame arcade in Moscow in the early '90s was stocked purely with Atari 8-bit machines!

Other unusual deployments for this chip include Epson's Japanese 486SX beast, the PC-486MU2GW, the Fujitsu FM-77AV family, and the non-MSX Sharp MZ-2500/2520 ("Super MZ").

Arcade Soldiers

However niche the PCs it first ended up in, the chip had much wider deployment in the arcades, not least as part of the SNK Alpha 68000 arcade platform (*Sky Soldiers, Time Soldiers*).

Classic Capcom game *Commando* had two of the chips controlled by an additional Z80.

Later, the chip became a minor partner with more powerful chips, such as the YM2151, the YM3812 OPL2 (as seen in Adlib cards), or OPL2's cut-down sibling, the YM2413.

This chip also sneaked into Sega's *Space Harrier* and *Hang-On*, and Taito's *The Legend of Kage* and *Arkanoid: Revenge of DOH*.

```
 1UP     TOP SCORE
  0        50000
```

COMMANDO

RANKING BEST 7

```
1ST    50000   VULGUS....
2ND    30000   SON.SON...
3RD    20000   HIGEMARU..
4TH    19420   CAPCOM....
5TH    12000   EXED.EXES.
6TH    10000   COMANDO...
7TH     8000   ..........
```

CAPCOM
COPYRIGHT 1985
ALL RIGHTS RESERVED

(1984) Ensoniq ES5503

Underused 32-voice wavetable/sample playback chip, crippled by Apple's design decisions with a happier life in professional synths.

As seen in: Apple IIGS, Ensoniq's Mirage, ESQ-1, and SQ-80.

32 mono/16 stereo 8-bit wavetable channels, 64k sample memory. Apple IIGS OS forces 15 stereo channels and reserves one channel for system sounds. Stereo output is deployed as mono on IIGS.

Ahead of its time

When Bob Yannes left Commodore, taking his vast synth knowledge with him, he realised his dreams at Ensoniq. This chip reflects the move away from using simple waveforms as musical building blocks, in favour of using samples. It also throws in a hint of the additive synthesis capabilities of AMY.

Even though this is the closest sequel to the SID chip, it's only the bells, whistles and raised standards that bear much resemblance: creeping into the small print are echoes of SID, such as the powerful filters, and "sync".

The chip has hidden depths. While it has 32 channels, they're internally grouped into eight "voices", which can modulate the frequency or volume of the other oscillators in esoteric ways.

Trappled

While this is an incredible chip (especially for 1984), the only home computer containing it (the Apple IIGS, launched in 1986) was not really a choice for budding audio creatives needing an all-in-one system, and the chip was somewhat trapped.

Two design decisions compromised the chip's deployment in the IIGS. A meagre 64K of sample RAM was assigned to it, and the headphone jack of the machine only outputs mono sound, despite the chip being stereo. In fact, stereo requires an extra expansion card, and sampling ability is an expensive third-party option.

So, while this chip is superior to Paula appeared in expensive synthesizers, it wasn't as useful or usable in the IIGS. Also, Commodore threw the technical gates open to their Amiga users, and Apple kept them almost shut, while charging a king's ransom for the privilege of opening them at all.

Some of the blame for this catalogue of audio issues falls on The Beatles' Apple Corp., which had extremely twitchy lawyers looking for any sign of Apple Computers entering the music business.

This caused Apple to be overly cautious about what features they exposed.

Underused

But what kind of soundtracks were released on the Apple IIGS? Did the composers take advantage of this 32-voice miracle chip? That would be a hard "no."

It seems that there are very few Apple IIGS soundtracks that trouble the 32-voice limit, with few even surpassing four-voice polyphony.

This is partially because most games are ports, starting off in the arcade (8 voices), Amiga (4 voices), or even the Apple II (beeper).

It might also be that the sample RAM has scant room to store many of the 8-bit samples needed.

The technical limitations also drag composers towards genres that are more suited to limited polyphony without sounding compromised. Dramatic soundtracks are out, sample-based, dancey tracker music is in.

Aside from samples, the chip is capable of esoteric synthesis techniques such as additive synthesis, but hardly anyone used those for commercial games during the lifespan of the system. This may be because of the extra development work needed to take full advantage of the chip.

There were, however, public domain tools originating from the tracker community, such as NoiseTracker GS v2.0 and Soundsmith, both of which focus on a tracking/sample-based approach.

No moos is good moos

The game *Ancient Glory*, a mythical horizontally scrolling walk-and-hit has some amusingly inappropriate sound effects. When you shoot a bull with an arrow, you get an orchestra hit! When you meet the very non-bovine end boss? You get mooing.

I haven't been so confused since the last Llamasoft game I played.

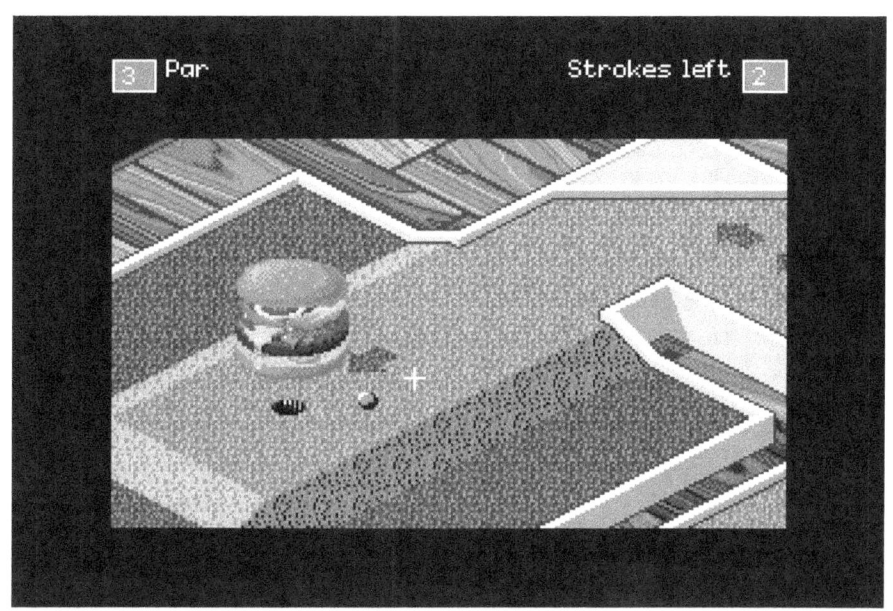

Zany Golf is zany.

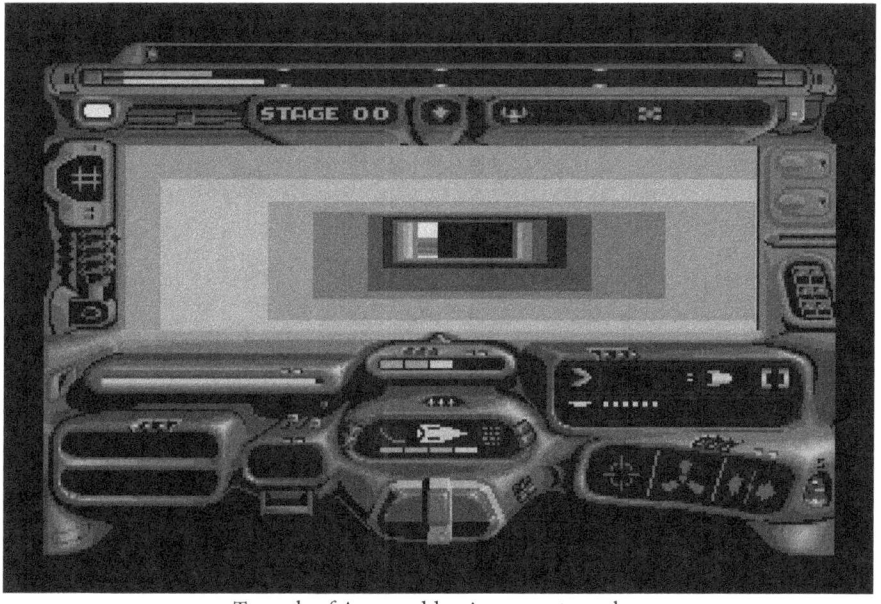

Tunnels of Armageddon is… er… tunnels.

(1984) Commodore MOS7360 (TED)

I'm not angry, I'm just disappointed.

As seen in: Commodore 16, Plus/4, 116

2 sucky square/noise channels over 4 terrible octaves, with no other features of note, except a barely adequate 10-bit pitch register. The noise generator is grindy too.

If the AY-3-891x was Coke, the SN76489 was Pepsi, and the SID is Inca Kola, then TED is New Coke.

In the 1980s, releasing a machine with just a beeper (ZX Spectrum, Acorn Atom, Sinclair QL) or no sound at all (ZX81, Amstrad PCW) was an actual commercial option. We're all grateful for companies that took the effort to be nice to the ears of their customers/players, no matter how flawed or limited the results.

But it's difficult to be grateful for the MOS7360 TED.

Found behind the sofa

Commodore lost SID's Bob Yannes when he left to form Ensoniq, and the TED chipset had been hanging around the company for some time. It was designed to be put into a cheap $49 business computer whose job was to fight a fearsome Japanese invasion of cheap home micros that never materialised.

That model, the C116, shipped into Europe in mid-1984 as a rubber-keyed machine only just larger than a ZX Spectrum. Its commercial failure was unsurprising.

The Commodore 16 (C16) that followed wasn't aimed at business like the C116. It was intended to replace the VIC-20, which was still in production.

It was the Plus/4 that was intended for business, having a larger memory than the C16 and (inadequate) business suites in the ROM. There was also a V364 variant, with a numeric keyboard and built-in voice synthesis.

Both the C16 and Plus/4 launched in late 1984 with TED on board and were a flop in the US. The V364 failed even harder. The Commodore 16 did reasonably well in Europe and Mexico, especially when it was later bundled with software (and after price cuts), but as a games platform it was unrewarding to program and often unrewarding to play. More colours and a better BASIC didn't help that much.

However, programmers such as Shaun Southern and Udo Gertz did sterling work, and companies such as Gremlin Graphics dutifully ported their games, with admittedly varying results.

Very few games were released that took advantage of the improved specification of the Plus/4, which was like the commercial neglect of the Commodore 128's enhanced features.

TED talk

As a sound generator, there's very little to say about TED. The only interesting thing about the chip is the corporate story behind why it exists.

It has two four-octave square wave generators, the second of which you can swap out for the noise generator.

Wikipedia also notes:

> "An undesirable feature of the chip is its well-known tendency to destroy itself through overheating."

On the bright side, it can be made to play samples with the usual oscillator-based methods, though whether it's a good idea to stress out a chip prone to overheating is an open question.

In a similar way to other SSG chips, switching the sound generators to constant level outputs and changing the volume register setting very quickly can turn a digital wave into an analog sound. Each generator can produce frequencies from 100 Hz to 23 kHz, which is pretty good considering how bad it could have been.

A narrow vision

Developers and enthusiasts soon delighted in taking the many limitations of the C16 and Plus/4 as a challenge. Decades later there are some great homebrew games and demos to finally give some comfort to anyone disappointed with how their 1984 Christmas turned out.

Maybe I was a tad harsh, though, because computer first love is a powerful thing, and it's perfectly possible to fall in love with limitations: sometime, the more crippling the better!

Writer and C64Audio.com research/PR guru Anna Black says this of TED:

> "I love the C16. My favourite computer ever. Probably childhood nostalgia, but there was something very good about the limitations as it forced you to think more about the code. I always remember Tony Kelly being a good C16 games programmer on Mastertronic.
>
> Mr Puniverse was epic!"

commodore 16

Owner's Guide

(1984) Namco CUS30

Powerful, stereo, and crunchy sample playback chip.

As seen in: Namco System 1, Thunder Ceptor and System 86.

8-voice, 4-bit wavetable @ max 24 kHz with no sample PROM needed. 20-bit frequency register.

A new Pac

Four years had passed since the Namco 15XX was rolled out on *Super Pac-Man*. Technology continued apace, not least because there was a new *Pac-Man* game to launch: *Pac-Land*, using the new Namco System 1 board.

In 1982 the 8 voices on the 15XX seemed like luxury, albeit with the same tiny wavetable samples as the original WSG.

This 1984 chip removed the need for a PROM, allowing as many wavetable samples as the game needed.

Further improvements added were a 24 kHz sample rate, and a massive 20-bit frequency register for all voices. Stereo also made an appearance, with each of the 8 voices pannable in stereo space.

Each channel was given the ability to take its output from a noise generator rather than a sample waveform. Powerful!

The arcade machines based on the *Pac-Land* hardware rely solely on the CUS30 for both sound and music. Other machines from *Thunder Ceptor* onwards use the YM2151 chip for music (as heard in *Marble Madness* and *Out Run*) and use this chip for SFX and some drums.

Pac-Land's isometric sibling *Pac-Mania* also used this chip, as well as *Galaga '88* and *Splatterhouse*, and it made an encore in *Pac-Man 25th Anniversary Edition*.

Too big, too long

Pac-Mania and *Galaga '88* also have two 8-bit DACs on board, allowing samples to be played straight from the CPU. But why do this when there's a perfectly good sound chip already there? One answer is that you can deploy samples that are better or longer than the CUS30 can handle (or which are encoded differently). However, there's a more subtle reason.

In *Galaga '88*, the DACs are used for playing extra samples such as sampled alien speech and a huge explosion whenever an enemy is hit.

The problem with that huge explosion is that the CUS30 allows sounds to interrupt other sounds, assigning available voices dynamically. This is great for making sure every sound is played.

However, if the long explosion was allowed to play every time you hit an enemy, you would end up hearing nothing but the start of the explosion machine-gunning on all 8 voices and drowning every other sound.

Outputting the sound directly under CPU control through a DAC, there's an occasional explosion that's played in full, with another one being played only when the first one has finished. This gives a frenetic feel without overwhelming the soundscape.

Pac-Mania uses the DAC to play applause, which Namco obviously didn't want to be cut off prematurely by other sounds in the game. The DAC is also used for the special item noise, so Namco must also have deemed it important enough to the gameplay to make sure the player heard it.

Splatterhouse uses the CUS30 for wing flaps and a high-pitched stressful whine when you're fighting snakes, but not much else. The DACs output thunder, groans, speech, and the main punch/shoot sound.

Rolling Thunder uses the CUS30 for dissolving enemies and square wave jingles. It uses a two-channel Namco 63701X DAC chip for bullets (channel 1) and speech (channel 2).

(1984) Yamaha YM3526 (OPL)

Influential FM chip behind a stack of home and arcade memories.

As seen in: Sound Expander (1985), MSX Audio (1986), Bubble Bobble.

2-op FM sound generator with mode selector enabling either nine melodic FM voices, or six melodic voices and five preset rhythm voices, most of which use 1-op plus channel-specific noise. Hardware vibrato oscillator and amplitude modulation. One voice has composite sinusoidal modelling possible (for speech synthesis).

The official datasheet for this 1984 chip tries to predict what it would be used for, with questionable accuracy:

> "The OPL (FM Operator Type-L) is a newly developed sound generator designed for CAPTAIN (Character and Pattern Telephone Access Information Network) systems and teletext."

This gives no clue to its ultimate fate, which was to power a few arcade machines and many FM add-on cartridges for home micros. Yamaha was always adept at repackaging and remarketing, and in this case, the OPL1 became the first in a trilogy of classic chips that eventually became the sound of PC Gaming for a generation.

Preset party

This chip varies from other FM chips up to this point because it contains some typically tinny percussion presets. These are FM, but with an element of generated noise.

Five sounds can be selected: bass drum, snare drum, high-hat cymbals, top cymbal, and tom-tom.

These presets are also in the later OPL2 and OPL3 chips in AdLib and Creative Labs cards.

The atmospheric tappiness of the Wolfenstein 3D music owes its feel to these Yamaha drum patches, which are economical to use, since you get five drums for the price of three voices.

One voice of the chip is also capable of voice synthesis, via "composite sinusoidal modelling" for replicating a singing voice, using roughly the same additive synth principle as used in AMY. This is what Earl Vickers did with the YM2151 chip in Atari's *Xybots*, but that's the only high-profile use of that feature.

Arcades and add-ons

The chip appeared in a startling number of arcade coin-ops, primarily from Data East, SNK and Nichibutsu.

Some of the more famous names are games which were converted to 8-bit micros such as *Terra Cresta*, *Bubble Bobble*, *Ikari Warriors*, *Athena*, *BreakThru* and *Renegade*.

While the game names are famous, the original composers are in some cases frustratingly anonymous!

It is as the designated chip in the MSX Audio standard that the chip is most remembered, though in this case the OPL core was integrated into a composite chip, the Yamaha Y8950.

The Y8950 has a built-in 8-bit PCM sample unit, 32 KB of sample RAM for ADPCM data, built-in mute, and filtering circuits, and general-purpose 4-bit input/output ports.

Looking at what happened with the Y8950 makes you wonder why anyone bothers with standards.

Only one cartridge for MSX1 was released that *actually* conformed to MSX Audio standards completely: the Panasonic FS-CA1. Other cartridges were released with partial compatibility, though these were also upgradeable to full compatibility (presumably at a price). Yet more wasted corporate meetings about standards hardly anyone used.

One interesting feature of the cartridges featuring MSX Audio is that they seamlessly add appropriate extensions to MSX BASIC. Some also feature music editors in firmware, along with friendly presets.

OPL in C64

This chip achieved its widest European audience as part of the legendary Commodore 64 FM Sound Expander. This cartridge could also work with its own full-size optional keyboard. For many C64 users, it was a chance to hear something new and exotic.

The demo software allowed sounds and rhythms to be selected and played. Most users ended up with the tape software which had fewer demo song choices than the rarer disk version.

Even harder to find were the optional *FM Composer* and *Sound Editor*.

The cartridge even had a passthrough port so you could use a sampler or even a Datel MIDI interface at the same time.

Hidden depths

For a consumer end-product, it is surprisingly upgradeable and programmable (even from BASIC) if you know what you're doing.

The OPL chip, for example, can be swapped out with an OPL2 with no other modifications.

Programming from assembly, the timing seems to have been a pain.

There are only two memory locations involved:
$df40 (register select) and $df50 (the selected register).

The program needs to wait 12 cycles between selecting a register using $df40 and that register being available at $df50. Though once a register is selected, it stays selected.

Convenience features in the demo software included the ability to play one-finger chords, a way to set the keyboard split, and the option to transpose the keyboard.

Fun fact:

The demo software doesn't check to see if the cartridge is there, it just writes to the relevant memory locations. This enables other software to hijack the pokes!

The modern FM-YAM cartridge replicates the Sound Expander and has even added new capabilities.

> *FM-YAM is a cartridge for your C64 that adds FM sound capability to your system. See it as adding Adlib to your C64, since it has the OPL2 YM3812 on board. Remember that a lot of 8-bit (and 16-bit) systems used to have Adlib options as well. There's the original Adlib sound cards for PC, but other systems, such as MSX computer peripherals, or Arcade games, as well as Yamaha synthesizers used these OPL2 chips.*
> **https://c64.xentax.com/index.php/fm-yam**

FM Expander – what does it offer?

Instruments
- Guitar
- Vibraphone
- Brass
- Strings
- E. Piano
- Organ 1
- Organ 2
- Harpsichord
- Flute
- Synth 1
- Synth 2
- Synth 3

Rhythms
- Pop 1
- Pop 2
- Rock 'n' Roll
- Reggae
- Disco 1
- Disco 2
- Country
- Bossa Nova
- Ballad
- Swing
- March
- Waltz

Extra Voice Bank (Disk version)
- Banjo
- Mellow
- Space Bell
- Strings 2
- Plucked
- Cosmic Wow
- Sweet Flute
- Alien
- Glock
- Synthbass
- E. Piano 2

Riffs and Demo Tunes
- Country Banjos (Riff)
- Fairy Dance (Song)

Extra Riffs and Songs (Disk version)
- Big Band (Riff)
- Pop (Riff)
- Disco (Riff)
- Winner Takes It All (Song)
- Telstar (Song)

(1985) Sega 315-5218 (SegaPCM)

Excellent sample playback chip that's doing more than it seems under the hood.

As seen in: Space Harrier, OutRun, After Burner, Hang-On.

16 channel PCM @ 31.25 kHz, 12-bit samples.
448 KB sound ROM, samples playable as instruments.

Pseudo-stereo (stereo channels are duplicated mono channels), but left and right levels are independently controllable.

SegaPCM as used in the Sega Space Harrier hardware was implemented almost completely with 7400 series TTL logic chips, becoming the single-chip 315-5218 for subsequent releases.

If you've ever been welcomed to the Fantasy Zone and did indeed Get Ready, then you've met this chip before. It might even have formed a large part of your wasted youth.

SegaPCM was used heavily by its AM2 division between 1985 and 1991, mainly in the Super Scaler series of high-end systems. Later, it was superseded by the stereo, CD-quality MultiPCM (YMW258-F) built by Yamaha.

The chip's job was to do the sonic heavy lifting that the YM2151 FM chip couldn't. For instance, in *OutRun*, all the sound effects such as the cheers and tyre screeching noises were well beyond the scope of the music chip, and you wouldn't really want to interrupt *that* music anyway!

It was also one of the first sample playback chips to be used to beef up YM2151 soundtracks.

After Burner's distorted guitar and chunky drums were from the SegaPCM, and tracks such as "Final Take Off" sound underpowered without them.

Multi-chip madness

The SegaPCM chip was created to fill a technical gap in an off-the-shelf solution, the YM2151. Many arcade games also found themselves needing more channels, richer sound, or access to a different type of sound such as speech.

The use of multiple sound chips requires more sophisticated drivers and greater creativity. Practical problems to be solved include timing, volume balance and synchronisation issues.

This illustrates one of the main differences between arcade coin-op machines and home micros/consoles: the budgets were bigger on coin-ops, so they could use multiple sound chips.

Many other companies installed off-the-shelf ADPCM chips from companies such as NEC to avoid the extra R&D required for custom silicon. These chips were less capable than SegaPCM, often being designed for speech and being monophonic.

SegaPCM games: that list in full!

A.B. Cop

After Burner/After Burner II

Enduro Racer (YM2151 version)

Enduro Racer (YM2203 version)

Galaxy Force 2

G-LOC Air Battle

G-LOC R360

GP Rider

Hang-On

Last Survivor

Limited Edition Hang-On

Line of Fire / Bakudan Yarou

OutRun

Power Drift

Racing Hero

Rail Chase

Space Harrier

Strike Fighter

Super Hang-On

Super Monaco GP

Thunder Blade

Turbo OutRun

(1985) Commodore MOS8364 (Paula)

Legendary sample playback chip with an almost religious following... and the hardest panning ever!

As seen in: Amiga family.

4-voice wave playback of 8-bit audio, at a maximum rate of 22,050 Hz. Uses DMA ("direct memory access"). Chip capable of amplitude and frequency modulation.

Consumer, meet sampler

Prominent Amiga composer Allister Brimble:

> "The sample playback was all hardware. Four channels of 8-bit samples and a filter that could be on or off. Interestingly, turning off the low pass filter also turned off the power LED!
>
> The first program to really make use of it was Aegis Sonix. It had sample playback and software synth as well. After that it was Soundtracker and Protracker all the way. This software could manipulate the samples via commands to adjust the volume, pitch, and timing of each note in various ways.
>
> Many composers who are now in the UK dance music charts started with Amiga sampling."

Tech overview

The Paula chip, designed by Glenn Keller, has 4-voice wave playback of 8-bit audio, at a maximum of 22,050 Hz. It uses DMA ("direct memory access") to read samples directly from chip memory. This avoids hogging the CPU and allows full use of the sound capabilities when there are other complex processes at work.

It has stereo output that consists of two voices hard-wired to the left stereo channel, and the other two to the right.

Notably, it can also use voices to modulate each other both by frequency (making FM-like instruments possible) and by amplitude (making sounds like ring-modulation possible). Both appeared in the early application Aegis Sonix, and in the occasional adventurous demo.

Looking back on the raw audio specs of the Paula from this millennium, it seems underpowered. However, that's only a small part of the story of this chip and the musicians that loved it.

A studio in the home

The initial A1000 was the first home computer with an integrated and usable sampling capability - beating the first affordable outboard sampler (Akai S900) to market. It could do a lot more than any studio box, being overall the most powerful home computer launched up to that time.

However, it took until 1987, when the Amiga 500 hit the UK and Europe at a relatively reasonable price (£399 in the UK, with a bundle of software), that the story of the cultural impact of the Amiga really starts. It was the Batman pack released for Christmas 1989 that really lit the fire:

> *"The fact is that the Batman pack sold 186,000 units from October to end December 1989"*
> **David Pleasance, former Commodore UK managing director.**

In 1987, Paula wasn't the most sophisticated sound chip in a home computer. That was the Apple IIGS, recently launched in the United States, which featured the 32-mono/16-stereo channel Ensoniq ES5503 designed by Bob Yannes, who also designed the SID.

However, to vast swathes of European youth, the Apple IIGS wasn't even an option - too expensive, too American, and boring.

The A500, in contrast, was relatively affordable (with some saving up), powerful and supported by a huge software catalogue of both commercial and shareware games.

It was also the gateway to a vibrant demo community. The Amiga provided an entry point to programming, creativity, and music - an all-in-one box that redefined what was creatively possible outside of dedicated studios.

A critical mass of creators meant that the number of tools, tricks and expansions grew very quickly. The ability of Paula to call for samples via interrupts was soon exploited to grow the apparent voice-count to 6, 7 or 8 voices, software mixed to defeat the hard-stereo panning.

Chipmusic that sounds like SID is popular on the Amiga. With the likes of *Octamed* musicians can replicate standard SSG waveforms by sampling a single-cycle and playing them looped like regular instruments. If you sample 8 pulse waves at different widths and play them sequentially, you have an emulated and fizzy pulse-width modulation.

This is the same wavetable technique seen in the Namco WSG but it is an artistic choice rather than a necessity.

Chip of the people

The Amiga's long lifespan created careers and community. Many musicians, graphicians and programmers made their names on this platform going on to long and illustrious careers, such as Barry Leitch, Allister Brimble, Andrew Barnabas, and Mike Clarke.

The backbone of the "demoscene" community continues to this day with lifelong friendships being celebrated yearly, and competitions to abuse the hardware to produce spectacular graphics and sound.

The actual audio Paula produced at the time may now sound like a degraded version of today's synths, but they're a sonic reminder of a youth when everything seemed possible.

In those days, boundaries were pushed every time you got a disk through the post, and that feeling is still potent to this day. The very human urge to create and remix never dies.

Paula Arcadia

On paper, bringing the home into the arcade with an Amiga-based gaming platform was a brilliant idea, but it never found success.

In 1986, Grand Products released the game *Up Scope* also based on Amiga hardware. The game is a supercharged version of Midway's *Sea Wolf II*, proving that good ideas are always worth recycling.

An Amiga 500 was used as the basis of the Arcadia Super Select board, featuring games such as *Sidewinder*, *RoadWars* and *Leader Board*. However, the results were a creative and commercial disappointment. Five years later in 1992, an A500 and laserdisc player were behind American Laser's game *Mad Dog McCree*, also powering the sequel. Additional games on this platform included *Space Pirates* and *Crime Patrol*.

(1985) DPC/Dave

Quirky custom chip partly responsible for massive delays in its host's launch.

As seen in: Enterprise 64/128

4-voice SSG. Stereo.

The story of the Elan Enterprise, which became the Flan Enterprise, which became the Enterprise 64/128, is one of ambition and delay.

First announced in 1983, the attractive design and integrated joystick of the machine turned heads in the computer press. Its custom graphic and sound chips gave it amazing specifications.

Unfortunately the computer was beset by problems. The first was its name, which they were initially forced to change to Flan Enterprise because of a conflict with another company named Elan.

Most of the rest of the delay was due to the complexity of the custom chips, which required extensive debugging. The computer limped into the market in 1985 as the Enterprise 64/128 duo.

Not just any old square wave generator

The custom sound chip provides three square-wave channels and a noise channel. Innovatively, the firmware allows envelopes with 255 phases/steps, each of which can also adjust the sound's pitch and stereo balance.

The design also takes inspiration from both SID and POKEY as well as anticipating features later implemented into systems such as the PC Engine.

The SID influence comes from voices being able to modulate each other, but in much more flexible ways. Unlike the SID, the ring modulation effect on a channel can be based upon the output of any other channel.

Equally, each channel can be distorted as on the POKEY, as the datasheet explains.

> "The 3 tone generators produce a square wave with frequency programmable from 30Hz to 125kHz which can be modified in various ways:
>
> Distortion can be introduced by using the output frequency to sample H.F. clocked polynomial counters. PN counters which can be selected are 4, 5 or 7 -bit. The 7-bit PN can also be exchanged for a variable length 17/15/11/9 -bit PN counter.
>
> A simple high pass filter is provided on each channel clocked by the output of a different channel.
>
> A ring modulator effect is provided on each channel, with the output of a different channel for its other input."

The noise channel is also much more flexible and configurable than other implementations, being able to feed off other channels as well as being fed through FX. As SSG noise channels go, it might be the most powerful ever designed.

The datasheet explains further:

> "The noise channel is normally a 17-bit PN counter clocked from 31kHz, generating a pseudo white noise. The input to this counter can be changed to clock of any of the 3 tone channels, and the PN counter can be reduced in length to 15, 11 or 9 bits. This counter can also be

> exchanged for the 7-bit PN counter. The resulting noise is then passed through high-pass and low-pass filters and a ring modulator, each controlled by the output of a different tone channel."

Finally, the two audio outputs (left and right) can both be turned into 6-bit DACs to output sampled sound.

By the time it arrived in the market, its competitors were a fraction of their launch price and boasted thousands of games. Additionally, 16-bit computers were up-and-coming.

As a result, it sold poorly, and there was only a limited amount of software released for the platform at the time. Inspired by Amstrad's "Amsoft", set up to help provide games for their CPC platform, Enterprise set up their own software house "Entersoft". As well as releasing disappointing original games that failed to make the most of the sound chip, they licenced games from other games companies. Games licenced from companies such as Loriciels and Mastertronic were released under different names to obscure their origin. Budget release *Finders Keepers* became full-price *King of the Castle*, and *Nonterraqueous* became *The Abyss*. These two games also featured the best soundtracks on the platform at the time, implementing music composed by legendary VGM composer Rob Hubbard, though not programmed by him.

As with all doomed and rare computers, dedicated fans (especially in Hungary where a quarter of the machines were shipped) have kept the machine alive with fan sites and good quality software and ports.

Only 80,000 computers were ever shipped, making it officially rare, as opposed to eBay "RARE".

(1986) Konami K051649/K005289 SCC1

Wavetable add-on for MSX cartridges that made it into the arcades.

As seen in: Nemesis 3 (MSX2), Bubble/GX4000 System Board.

5-channel wavetable playback with 32-byte samples (signed 8-bit). 12-bit frequency register, 128-byte sample memory in total.

This chip shares wave memory between voices 4 and 5.

There is an improved K052539 variant nicknamed "SCC+" in a sound cartridge delivered with the disk games Snatcher and SD Snatcher. This upgrade gives voices 4 and 5 their own slots.

POWERUP

The Konami SCC1 started life as a hardware upgrade built into some MSX cartridges, to complement the existing sound chip (AY-3-8910 for MSX1, YM2149 for MSX2).

The volume levels between the two chips were designed to be compatible, which was one less thing for the music programmer to worry about.

32 Bytes of Wavetable Wonder

On the Commodore Amiga, some trackers chose to emulate the PSG chiptune aesthetic on the Paula chip by deploying tiny samples (which were single-cycle or waveforms).

What is a stylistic choice with Paula is a necessity on the SCC1, which is resolutely wavetable-based.

There are five voices, and four 32-byte slots (channels 4 and 5 share a waveform). These slots contain tiny 32-byte samples, used as the basis for both looped melodic instruments and single-hit sounds and percussion.

This was first implemented on the Namco WSG, but the resolution and quality is better on the SCC1 (8-bit SCC1 versus 4-bit WSG).

There are some types of sounds a small wavetable chip is unsuitable for, but composers could still rely on the existing AY-3-8910 or YM2149F chips in the target machine for smooth SSG leads, wibbled chords or noise-based effects.

This chip was installed into:

- *Parodius* (Japan only, the series originated on the MSX1)
- *Salamander* (Japan, Europe, MSX1)
- *Contra/Gryzor* (MSX2)
- *Space Manbow* (MSX2)

SCC1 hits the arcade

This chip was also designed into Konami arcade coin-ops such as *Haunted Castle*, *City Bomber* and *Slime Kun*.

Slime Kun has only a NEC μPD7759 speech chip to keep it company, so the music is pure SCC1.

Haunted Castle has an added YM3812 OPL2.

The arcade variant of this chip was designated K005289. It was installed into Namco's short-lived Bubble System, and its subsequent ROM system, being responsible for the sound effects in *Gradius*, helping two AY-3-8910A chips, four filter chips and a Sanyo VLM5030 speech synthesizer.

Super Tech Talk: Noise Generation in 8-bit Sound Chips

Many 8-bit sound chips provide a waveform called "noise". It was often used for percussion and sound effects.

The exact algorithms used to generate the noise waveform were different for each chip, but they had some basic principles in common.

The noise channel was created from some sort of randomness. Modern chips can generate truly random noise, but 1980s chips, with limited die space, could not. Instead, they used pseudo-random noise, which sounds a lot like noise to our ears but is mathematically quite different. To generate this noise, chip designers used what were then called polynomial counters, known today as Linear Feedback Shift Registers (LFSRs). LFSRs were used for all kinds of randomness in 8-bit computing because they can easily be implemented in hardware or software, and they have provable properties.

Before we get into those properties, here is a quick overview of how an LFSR works.

A shift register is an array of flip-flops (one-bit memory cells) that moves a bit down the line on each clock cycle. If you have a shift register length of 16, you can put bits into the register, and they will be output at the other end after 16 clocks, in first-in, first-out (FIFO) order. An LFSR is a shift register in which the next input bit depends on some combination of the bits already in the pipeline. That's the feedback part. The changes are implemented via exclusive-or (XOR) operations, which are bitwise addition without carries. That's the linear part. The bits that feed back into the input are known as taps.

An example 16-bit LFSR is shown below.

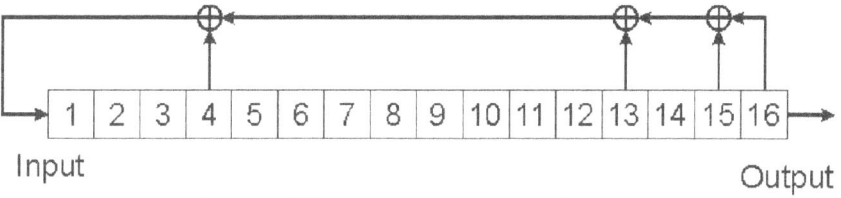

A 16-bit LFSR with taps at bits 16, 15, 13, and 4,
designed to have the maximum period of 65,535.

The LFSR is loaded with an initial state and clocked. It generates its own inputs from the existing bits in the pipeline. It might not be immediately obvious, but if you load an LFSR with all zeroes, it stays that way forever. Usually, they are initialized to all ones to prevent this.

At some point, an LFSR will end up back in the same state as it started, and it will repeat the same sequence of bits again and again, forever. The length of the bit sequence up until the state repeats is called the LFSR's period. For an n-bit binary LFSR, the period must always evenly divide $2^n - 1$, which is also conveniently the number of n-bit numbers (minus one, because the LFSR can't output zero).

For every n, there are many full-period LFSRs with period $2^n - 1$, and it is those LFSRs that are most suitable for generating pseudorandom sequences.

As an interesting aside, if you pick n such that $2^n - 1$ is a prime number, then all LSFRs will have the maximal period! In that case, you can build a maximal-period LFSR with only one tap, which saves transistors. A particularly convenient n with this property is 17, which was used in several sound chips (including the POKEY and SID); it can be built in hardware with 17 flip-flops and a single XOR gate and has a period of 131,071.

Another interesting feature of maximal-period LFSRs is that they visit every possible number between 1 and $2^n - 1$ exactly once during each period, but in a random-looking order. So they can be used for effects such as screen dissolves, where every pixel must be visited once but you want to visit them in a random order.

One design choice with LFSRs is how to sample them. You could either drive them from the system's clock (which would cycle the LFSR very quickly), collecting samples at the lower playback frequency. Or you could just clock the LFSR each time a sample was needed. Nearly all sound chip LFSRs had periods with prime number lengths, so it was easy to sample the entire set uniformly, even when skipping states due to the different clock/sample rates.

The bits from the LFSR were either output raw or sent through a DAC for the noise channel.

Pitched Noise

Many 8-bit sound chips support *pitched noise*. That is, when you specify a noise sound, you can also specify a pitch (or note) for that noise. Noise sources are frequently used for 8-bit percussion sounds, but pitched noise is not the same thing as perceived pitch from percussion instruments such as tom-toms or timpani. Those instruments really do create a pitch and our perception of it works in pretty much the same way as it does for most music.

First, a quick overview of noise. Most audible noise is primarily either *white noise* or *pink noise*. White noise has equal spectral power at all frequencies and sounds like a hiss. Computers and digital electronics make it simple to create white noise. Pink noise has a power spectrum that falls off as the frequency increases and sounds more like the ocean or a waterfall. Most of the noise sounds we hear in the natural world are pink noise.

The best way to illustrate the various noise types is by looking at their frequency spectra. A frequency spectrum shows the energy at each frequency (specified in cycles per second, or Hz), using a logarithmic dB loudness scale. White noise has a flat spectrum with equal energy at all frequencies, as shown here:

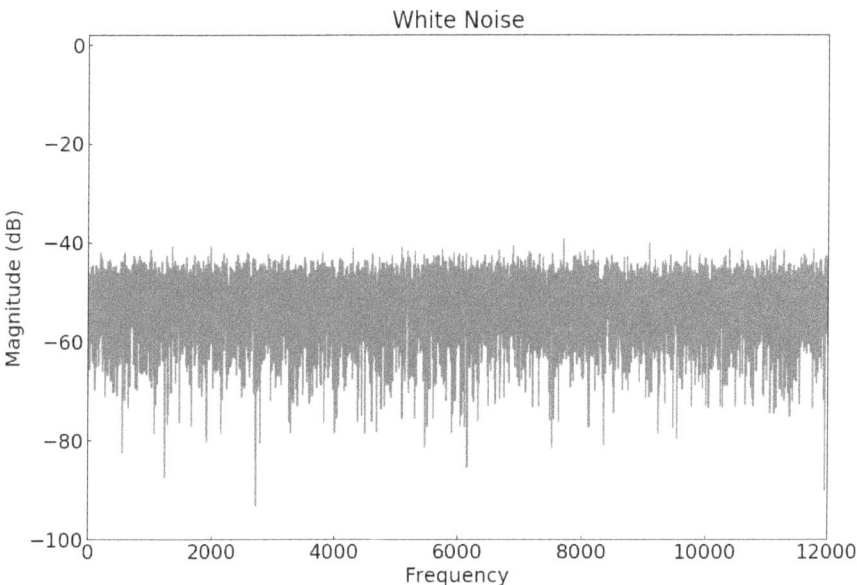

Analyzing the spectrum of a noise source is very helpful to understand the characteristics of the noise. So, for the chips discussed here, I will also show the power spectra.

Pink noise has more of its energy at low frequencies. The spectral power falls off as 1/f as the frequency increases, as shown here:

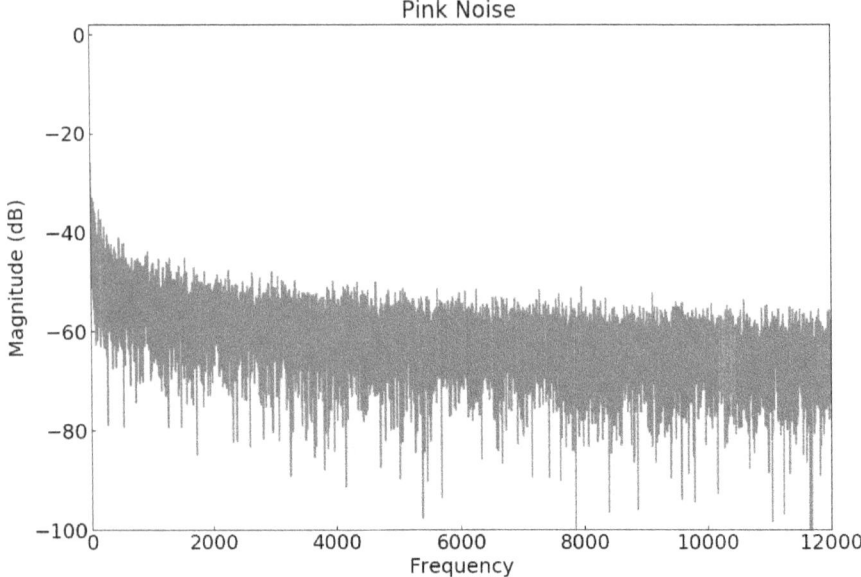

Despite what you may have heard, computers do not naturally create either white noise or pink noise. That said, most computer-generated noise is nominally intended to be white. Computers can create uniform random numbers, which is one requirement for white noise, but they can only do so at discrete times derived from a system clock, which does not have infinite time resolution. True white noise requires both uniform random numbers and random times. If the computer updates the random numbers quickly enough, though, the noise at much lower frequencies can appear very much like white noise. For audible frequencies (up to 20 kHz), uniformly distributed random numbers sampled at frequencies of a few hundred kHz would be indistinguishable from true white noise.

It would be impossible (and extremely boring) to go into technical detail about the noise from every sound chip, so we'll just look at a few representative examples. For these examples, the noise was programmed with a pitch of 440 Hz, which is a standard frequency used for tuning. The chip that can produce sound closest to white noise is the General Instrument AY-3-8910. Its noise spectrum in the audio range is almost identical to the pure white noise shown previously, but available information about how its noise generator works is difficult to find. Based on the spectrum, though, it must have been sampling the noise at a substantial fraction of its internal frequency of 1 MHz.

A more common chip, the Gameboy DMG, could produce sound relatively close to white noise:

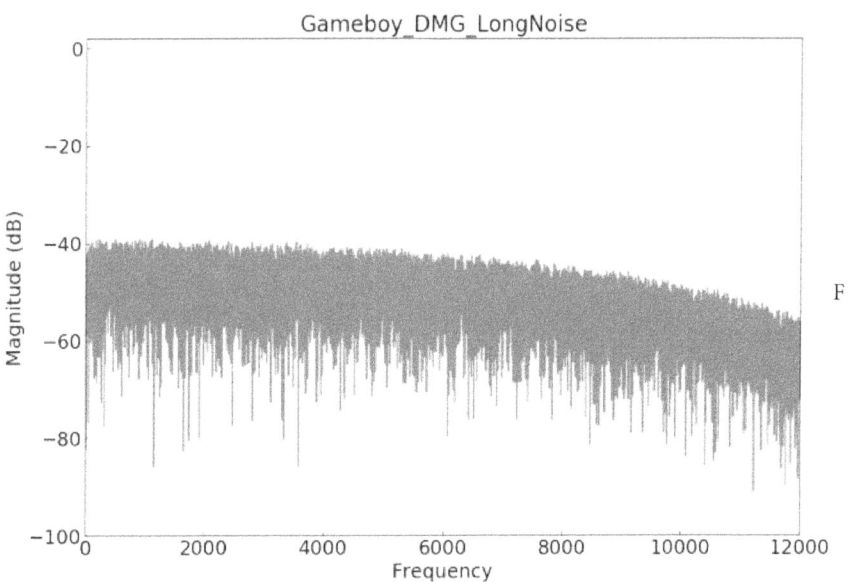

Spectrum for Gameboy DMG "long noise." Nominal pitch = 440 Hz.

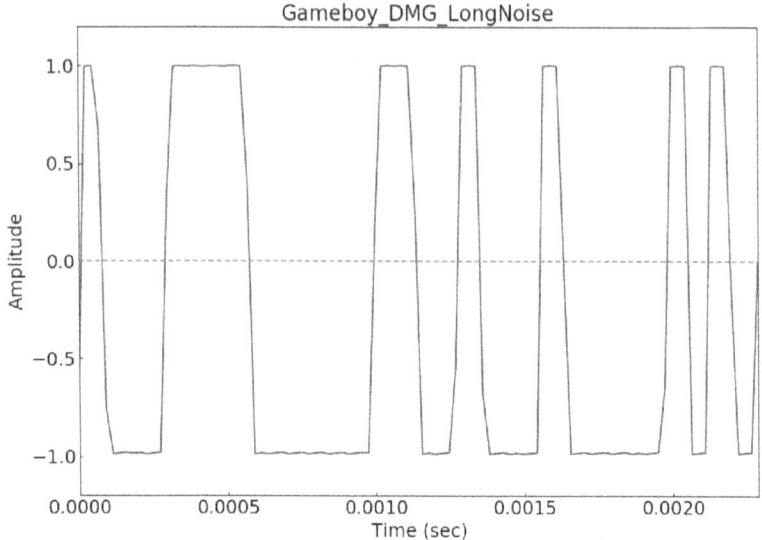

Waveform for Gameboy DMG "long noise."
Nominal pitch = 440 Hz.

So below the spectrum is the waveform for a single cycle at the programmed pitch. The noise was generated by randomly setting the output to either a 1 or a 0 a total of 32 times during each period. This chip is a good example of what is called 1-bit noise.

1-bit Noise

Many chips generate 1-bit noise, but at considerably slower bit rates. When you look at one of those (the SN76489) the big picture becomes clearer:

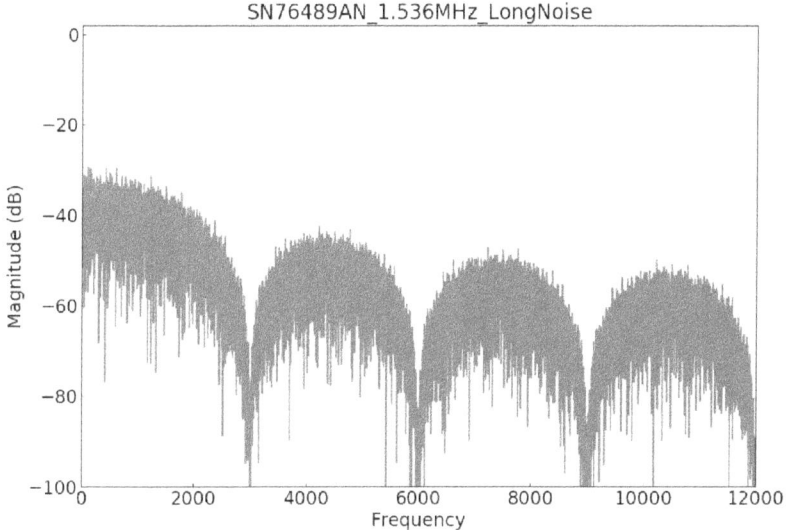

Spectrum for SN76489 "long noise" at a clock rate of 1.536 MHz.
Nominal pitch = 440 Hz

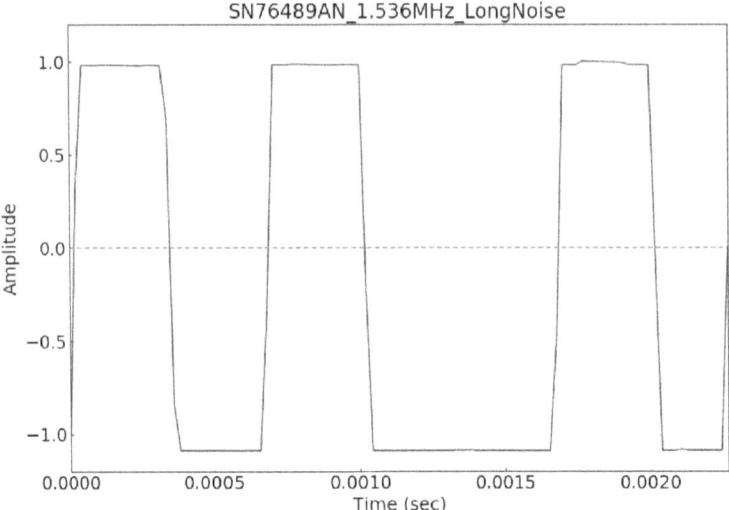

Waveform for SN76489 "long noise" at a clock rate of 1.536 MHz. Nominal pitch = 440 Hz.

The 1-bit noise has big notches in the spectrum at multiples of the bit rate. What's interesting in this spectrum is that the pitch you hear is fairly near the position of the notch! Usually, in audio spectra, the pitch you hear comes from a peak, with more energy at the frequency. But here there is less energy, yet your brain still perceives a pitch.

Any noise generation algorithm that produces noise by changing values at discrete times will have these notches. Even the AY-3-8910, which made very uniform white noise in the audio range, probably had a notch at a very high frequency (so high that you can't see the effect in the audible spectrum).

So, for the most part, any pitch you hear in a digitally produced noise spectrum from the 8-bit era will be from spectral notches.

It turns out that there has been a lot of research about pitch perception in noise since the 1980s!

Researchers report that for a noise edge, where noise goes up to a certain frequency and then drops to zero, people hear a pitch 2-5% below the frequency of the edge. William M. Hartmann, in his 2019 paper "Monaural edge pitch and models of pitch perception" (The Journal of the Acoustical Society of America 145, 1813) found that for an edge with noise at frequencies above, people hear a pitch 2-5% higher than the edge.

So for a notch like this, somehow your brain adds all the information together to give a (rather weak) sense of pitch at the notch frequency. Notice that the notch occurs at the update rate for the noise, which in this case is 440 * 7 = 3080 Hz.

Still, your brain reconstructs the fundamental of 440 Hz from the overtones it hears.

One more thing to notice: this noise falls off with frequency, making it sound more like pink noise than white noise and giving it a more natural (and, to many, more pleasing) sound.

SID Noise

The 6581/8580 SID chips in the Commodore 64 were much more sophisticated. The chip has a digital-to-analog converter (DAC) that can output 4096 voltages, which means that it can produce much more complex waveforms. In the SID, the noise is created by sampling random 12-bit numbers at 16 times the frequency of the note value, which gives it a remarkably complex noise spectrum:

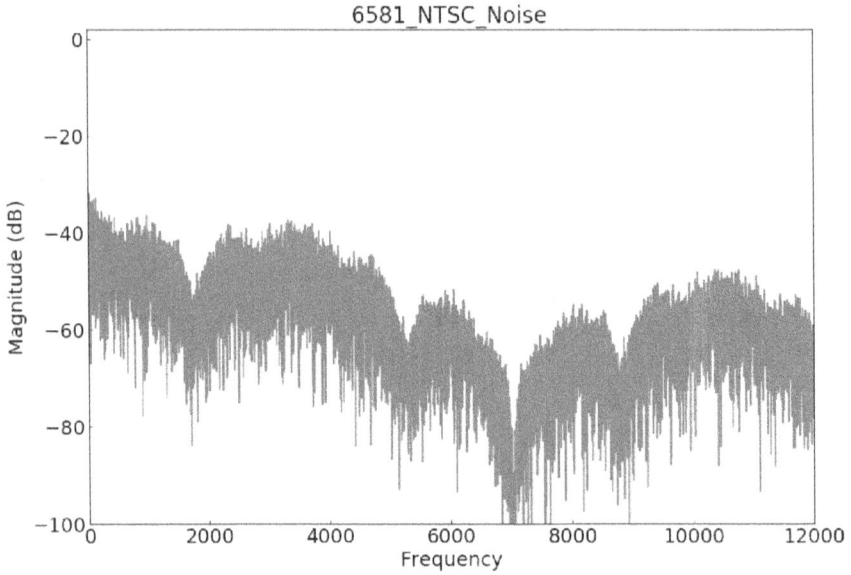

Spectrum for MOS 6581 SID noise. Nominal pitch = 440 Hz.

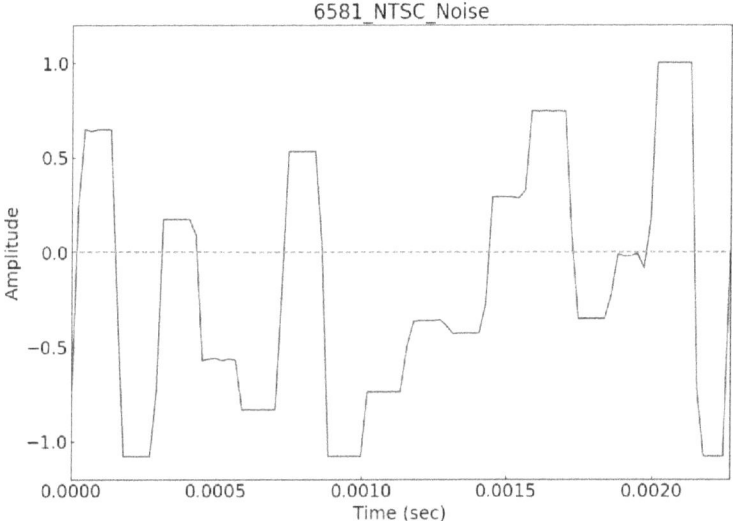

Waveform for MOS 6581 SID noise. Nominal pitch = 440 Hz.

While the primary notch in the spectrum is at the sampling frequency, the spectrum is much more complex. But that complexity was not necessarily intentional since it's a side effect of the algorithm used to produce the noise.

Some people believe that it gives a clearer pitch than the 1-bit sounds.

VIC-20 Noise

An example of a vintage system that made noise with a discernible pitch is the VIC-20, which uses the 6560/6561 chips for both sound and video. While the noise channel is described in terms that are like other systems, the output is very different. You can quite clearly hear the intended pitch. The spectrum and waveform are both dramatically different from other systems:

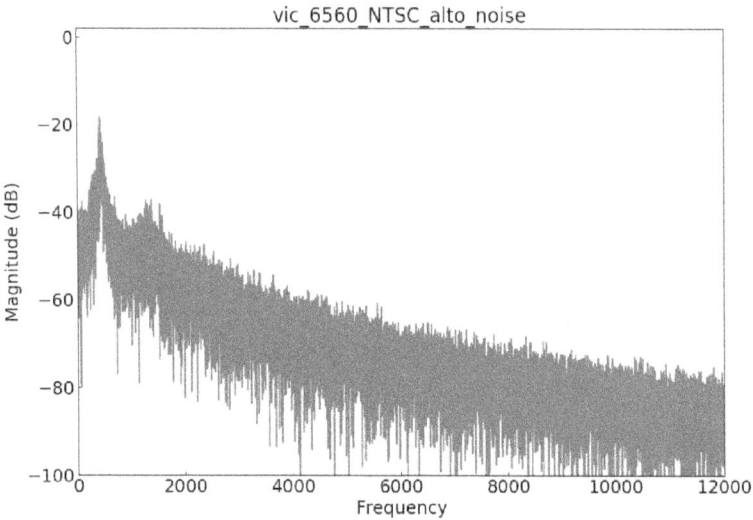

Spectrum for VIC 6560 alto noise. Nominal pitch = 440 Hz.

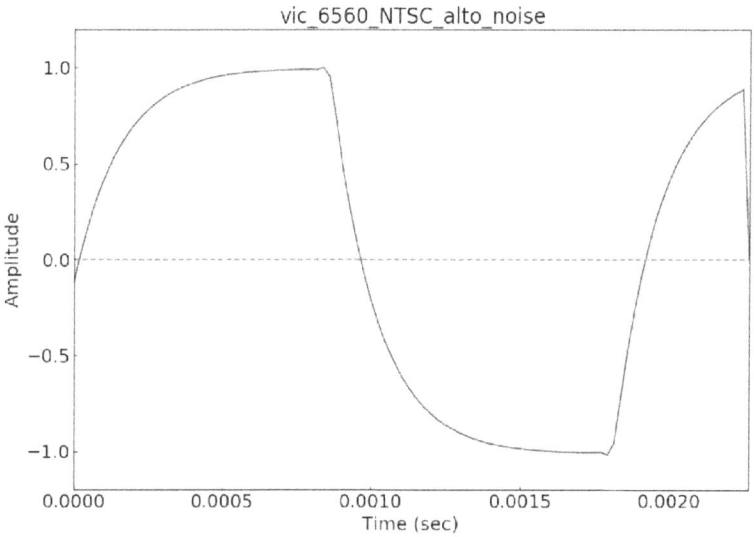

Waveform for VIC 6560 alto noise. Nominal pitch = 440 Hz.

The most obvious feature of the spectrum is the large peak centered at the chosen frequency. The waveform is dramatically different from the other chips for two reasons. First, the sampling period appears to be close to the nominal frequency, and second, the square waves have a distinct rounded shape (caused by capacitance in the electronics). Despite what you might think, the rounded shape has nothing to do with the perceived pitch! It just has the effect of damping the higher frequencies, which is apparent from the relatively steep dropoff of the spectrum. The peak in the spectrum, however, is a result of what appears to be frequency modulation; instead of the random data generating ones and zeroes at a high frequency, it appears to modulate the base frequency by the random data. That makes the pronounced peak in the spectrum, and the clearly audible pitch to the noise.

Super Tech Talk: Assigning pitches to sound chips

by David Youd

Some chips produce super-accurate notes. Some don't. Here's why.

Early sound chips each had their own numbering systems for the notes they could play. These note numbers would be written to and stored in the sound chip's registers, often by bedroom coders entering BASIC POKE commands.

Many numbering schemes exist for pitches, the most widespread of which is the MIDI note numbering system. A standard piano has 88 keys, from MIDI note 21 (A0) to MIDI note 108 (C8), with middle C (C4) having MIDI note number 60. The sound frequency for a MIDI note number *n* is given by:

$$f = 440 \times 2^{\frac{n-69}{12}}$$

8 and 16-bit era sound chips are not MIDI-based and don't use MIDI numbers to specify pitches; rather, each chip was designed with a pitch numbering scheme appropriate to its capabilities.

The mapping from these numbers to pitches could also be affected by the region the computer was designed for. Why, you ask?

Well, it dates to the 1950's, when countries' television standards diverged. America and Japan adopted the NTSC standard, and Europe adopted PAL. TVs tied their screen-draw rates to the AC power frequency: 60 Hz for NTSC and 50 Hz for PAL.

To synchronize CPU processing more easily with AC-synced TV/monitor screen animations, early home computers also tied their processor's clock speed to the regional AC power frequency.

The clock frequency of a chip can be given as the count of how many pulses the chip can generate in a second. For an NTSC Commodore 64, this is 1,022,727/sec (and 985,248/sec for PAL).

Sound chips have various ways of dividing this clock to create different pitches.

The C64's SID chip specifies a pitch n using 16-bit values (0 to 65535), which map onto a sound frequency according to this formula:

$$f = \frac{n \times clock}{2^{24}}$$

Example: A C0 (below human hearing) is 16.35 Hz. On the SID, the closest you can get with an NTSC machine is value 268 (16.34 Hz); for PAL, it's 278 (16.33 Hz). An A#7 is 3729.31 Hz, and with an NTSC SID, 61177 gives 3729.31 Hz, while 63504 on a PAL gives 3729.30 Hz.

Recall that pitch is on a logarithmic scale, where each higher octave doubles the frequency. This means that the frequency range for low notes is narrow, while the frequency range for high notes is wide. Given the relationship between SID note values and their frequencies, the SID has more 16-bit note values falling into the higher note ranges, meaning finer tuning control is available for the higher notes.

Given that the frequency width of notes varies, to make comparing pitches easier, the notion of ¢ ("cents") comes to the rescue. The frequency width of a tuned (0¢) note n is always -50¢ < n < 50¢ (e.g. 49¢ is a very sharp note). Music written for a PAL system behaves differently on the faster NTSC machine, with 19% faster playback and at a 65¢ higher pitch. Another interesting consequence of the formula is that a B7 can be played on a NTSC C64 (64815), but not on a PAL since the value 67280 can't be expressed in 16-bits.

In short, the SID has fine control over pitch. Less-capable sound chips must select from a much smaller set of frequencies. To illustrate, the Commodore VIC-20 only has 7-bit pitch values, which creates more coarse clock divisions. The VIC-20 NTSC clock is 1,022,727 (like the C64), and the PAL clock is 1,108,404. To work around the coarse division, each of the three square-wave channels have different, overlapping ranges.

These are achieved by a range factor r to further divide the clock: 256 for bass, 128 for alto, and 64 for soprano. The VIC-20 assigns a 7-bit note value n to a frequency as follows:

$$f = \frac{clock}{r(127-n)}$$

By having the 7-bit note value in the denominator, more note "POKE values" fall within the narrow-range low frequency notes than the wider-range high frequency notes. (This is the opposite of the C64).

This means the VIC-20 provides more precision in the low notes than the high notes. As the 7-bit note values increase for higher pitches, the high notes become overly sharp or flat, and, ultimately, not contiguously available.

For example, in the soprano voice, the 7-bit note value 125 gives a B8 (7,900 Hz), while the next note value (126) jumps all the way to B9 (15,980 Hz), skipping all the pitches in between.

While this is certainly limiting, some sound chips have an even more coarse clock divider, such as the 5-bit frequency divider in the Atari 2600's TIA chip.

Trying to adapt a known song to that platform is like navigating a pitch minefield.

www.ingramcontent.com/pod-product-compliance
Lightning Source LLC
Chambersburg PA
CBHW071355210526
45465CB00001B/93